STUDENT SOLUTIONS MANUAL

JOSEPH D. AUGSPURGER
GROVE CITY COLLEGE

PHYSICAL CHEMISTRY

A MODERN INTRODUCTION

CLIFFORD E. DYKSTRA

PRENTICE HALL, UPPER SADDLE RIVER, NJ 07458

Senior Editor: ***John Challice***
Production Editor: ***Dawn Blayer***
Special Projects Manager: ***Barbara A. Murray***
Supplement Cover Manager: ***Paul Gourhan***
Manufacturing Buyer: ***Ben Smith***

Printed in the United States of America

10 9 8 7 6 5 4 3 2 1

ISBN 0-13-604976-1

Prentice-Hall International (UK) Limited, *London*
Prentice-Hall of Australia Pty. Limited, *Sydney*
Prentice-Hall Canada Inc., *Toronto*
Prentice Hall Hispanoamericana, S.A., *Mexico*
Prentice-Hall of India Private Limited, *New Delhi*
Prentice-Hall of Japan, Inc., *Tokyo*
Simon & Schuster Asia Pte. Ltd., *Singapore*
Editoria Prentice-Hall do Brasil, *Ltda., Rio de Janero*

To Laura, Carolyn, Michele, and Michael

About the Author

Joseph D. Augspurger is a professor of chemistry at Grove City College. He holds a B.S. in Chemical Engineering and a Ph.D. in Chemistry from the University of Illinois at Urbana–Champaign. He was a Special Fellow of the Leukemia Society of America during a post-doctoral appointment at Cornell University. He has published over 30 papers concerning the physical chemistry of intermolecular interactions, applying both theoretical and experimental approaches.

Contents

Preface

This manual provides completely worked solutions to all exercises and additional exercises found in *Physical Chemistry: A Modern Introduction* by Clifford E. Dykstra. To facilitate their use, fundamental constants and atomic masses of the first 35 elements have been provided on the endpages, as well as two tables of conversion factors from Appendix VI in the text. References to equations in the main text are in brackets.

To the student: it should go without saying that you will gain the maximum benefit from these solutions by using them only after you have first made every effort to solve the problem yourself. Telling yourself "I know how to do this – I'll just make sure" is the quickest route to a failing examination grade. Physical chemistry is perhaps the most demanding subject you will face as an undergraduate; do not be surprised if it requires a great effort. But your reward will be a glimpse of the fundamental principles that will guide you in whatever area of chemistry you choose to pursue.

Bridget Dibble, DePauw University, and Clifford E. Dykstra, Indiana University–Purdue University Indianapolis, reviewed these solutions and made many helpful suggestions based on their personal experience. I thank John Challice, editor, for allowing me to be a part of this project and his help throughout. I also thank Barbara Murray, special projects manager, and Dawn Blayer, production editor, for finishing touches on the manuscript.

Finally, thanks to my family, who patiently put up with a very busy husband and father for the past three months.

Any comments you have regarding these solutions will be gratefully received at jdaugspurger@gcc.edu.

Joe Augspurger

Chapter 1

The World of Atoms and Molecules

Exercises

1. $E_J = B\,J\,(J + 1)$ [1-3]; $B = 3.836 \times 10^{-23}$ J.
   $E_0 = B \times 0 \times 1 = 0.$
   $E_1 = B \times 1 \times 2 = 2B = 7.672 \times 10^{-23}$ J.
   $E_2 = B \times 2 \times 3 = 6B = 2.302 \times 10^{-22}$ J.

   $g_J = 2J + 1.$ (see paragraph following [1-3])
   $g_0 = (2 \times 0) + 1 = 1.$
   $g_1 = (2 \times 1) + 1 = 3.$
   $g_2 = (2 \times 2) + 1 = 5.$

2. The state of 2 6s and 4 5s is made up of the following 15 configurations (or microstates):

[6,6,5,5,5,5]	[5,6,6,5,5,5]	[5,5,6,5,6,5]
[6,5,6,5,5,5]	[5,6,5,6,5,5]	[5,5,6,5,5,6]
[6,5,5,6,5,5]	[5,6,5,5,6,5]	[5,5,5,6,6,5]
[6,5,5,5,6,5]	[5,6,5,5,5,6]	[5,5,5,6,5,6]
[6,5,5,5,5,6]	[5,5,6,6,5,5]	[5,5,5,5,6,6]

3. Solution of this problem requires two steps; first, the determination of the total number of arrangements (configurations) of four dice and, second, the number of configurations in which three dice have one number and the

other die is different. First, since each die has 6 possibilities, the total number of configurations is $6 \times 6 \times 6 \times 6 = 6^4 = 1296$. To determine the number of configurations in which three dice have the same number and the last die has another, we first calculate the number of *states* consistent with this distribution. One such state is three 1s and one 2; however the lone die could also be a 3, 4, 5, or 6. Thus there are 5 possible states with three 1s. Because states with three 2s, 3s, 4s, 5s, or 6s are also allowed, the total number of states is $6 \times 5 = 30$. For each state, the number of configurations is [1-5]

$$\frac{4!}{3!1!} = 4.$$

Thus the total number of configurations (arrangements) of four dice in which three are the same and the fourth is different is $30 \times 4 = 120$. The probability of such a roll of the dice is $120 / 1296 = 0.09259$.

4. As in the text (p. 13), the original box whose sides are of length l is divided into smaller boxes whose sides are of length l/N; the total number of smaller boxes is N^3. The total volume is simply l^3. If the original cube is extended from l to $2l$ on one side, the total number of smaller boxes is now $2N \times N \times N = 2N^3$. The total volume of the expanded cube is $2l \times l \times l = 2l^3$. If the number of configurations is identified with the number of smaller boxes, it is true that

$$\frac{C_{\text{expanded}}}{C_{\text{original}}} = \frac{2N^3}{N^3} = 2 = \frac{2l^3}{l^3} = \frac{V_{\text{expanded}}}{V_{\text{original}}}.$$

Thus, $\Delta S = k \ln \frac{C_{\text{expanded}}}{C_{\text{original}}} = k \ln \frac{V_{\text{expanded}}}{V_{\text{original}}}$, as in [1-7].

5. $$\frac{P_2}{P_1} = \frac{e^{-E_2/kT}}{e^{-E_1/kT}} = e^{\frac{-(E_2-E_1)}{kT}} = e^{\frac{-(9.0-3.0)\times 10^{-21}(\text{J})}{1.381\times 10^{-21}(\text{J/K})\times 100(\text{K})}} = 0.0129.$$

6.

N	N ln N – N	N ln N – N + 1/2 ln(2πN)	Relative difference
10	13.0258	15.0961	13.7%
1000	5907.75	5912.13	7.41×10^{-4}%
100,000	1,051,293	1,051,299	5.71×10^{-6}%

Note how the relative difference becomes insignificant as N becomes large. Since this approximation will typically be invoked when N is on the order of Avogadro's number (6.022×10^{23}), the errors involved with making the approximation will be insignificant.

7. By definition, 1 mol of ^{12}C atoms is assigned a mass equal to 12 g. Since one ^{12}C atom has also been assigned the mass of 12 amu (*atomic mass units*), it follows that

$$12 \text{ amu} = 12 \text{ g mol}^{-1} / 6.0221 \times 10^{23} \text{ (atoms mol}^{-1}\text{)},$$

and therefore

$$1 \text{ amu} = 1.6606 \times 10^{-24} \text{ g} = 1.6606 \times 10^{-27} \text{ kg}.$$

He: $4.0026 \text{ amu} \times 1.6606 \times 10^{-27} \text{ kg} = 6.6467 \times 10^{-27}$ kg.
^{20}Ne: $19.9924 \text{ amu} \times 1.6606 \times 10^{-27} \text{ kg} = 3.3199 \times 10^{-26}$ kg.
^{40}Ar: $39.9624 \text{ amu} \times 1.6606 \times 10^{-27} \text{ kg} = 6.6362 \times 10^{-26}$ kg.

10 K: $3kT = (3)(1.3807 \times 10^{-23} \text{ J K}^{-1})(10 \text{ K}) = 4.1421 \times 10^{-22}$ J
300 K: $3kT = (3)(1.3807 \times 10^{-23} \text{ J K}^{-1})(300 \text{ K}) = 1.2426 \times 10^{-20}$ J
500 K: $3kT = (3)(1.3807 \times 10^{-23} \text{ J K}^{-1})(500 \text{ K}) = 2.0711 \times 10^{-20}$ J

We can now apply [1-42] to find $\langle s^2\rangle^{1/2}$. For He at 10 K:

$$\left\langle s^2 \right\rangle^{1/2} = \sqrt{\frac{3kT}{m}} = \sqrt{\frac{4.1421 \times 10^{-22} \text{ kg m}^2\text{s}^{-2}}{6.6467 \times 10^{-27} \text{ kg}}} = \sqrt{6.2318 \times 10^4 \text{ m}^2\text{s}^{-2}}$$

$$\left\langle s^2 \right\rangle^{1/2} = 249.6 \text{ m s}^{-1}$$

$\langle s^2\rangle^{1/2}$	10 K	300 K	500 K
He	249.6 m s^{-1}	1367.3 m s^{-1}	1765.2 m s^{-1}
^{20}Ne	111.7 m s^{-1}	611.8 m s^{-1}	789.8 m s^{-1}
^{40}Ar	79.0 m s^{-1}	432.7 m s^{-1}	558.6 m s^{-1}

8. $m = 3kT/\langle s^2 \rangle$

$= 3 \times (1.381 \times 10^{-23}\ \text{J K}^{-1}) \times 300\ \text{K} / (350\ \text{m s}^{-1})^2$

$= 1.014 \times 10^{-25}$ kg.

$1.014 \times 10^{-25}\ \text{kg} / 1.66 \times 10^{-27}\ \text{kg amu}^{-1} = 61.09$ amu.

Additional Exercises

9. There are three possible types of states: all three dice have the same number, two are the same and one die is different, or all three are different. Now, as in Problem 3, the number of states of each type must be determined, as well as the number of configurations for each state. Using the notation introduced in the text in Section 1.4, there are four states in which all three dice have the same number: (3,0,0,0), (0,3,0,0), (0,0,3,0), and (0,0,0,3). For the second type, there are 12 states: (2,1,0,0), (2,0,1,0), (2,0,0,1), (1,2,0,0), (0,2,1,0), (0,2,0,1), (1,0,2,0), (0,1,2,0), (0,0,2,1), (1,0,0,2), (0,1,0,2), and (0,0,1,2). Finally, there are four states where all three dice are different: (1,1,1,0), (1,1,0,1), (1,0,1,1), and (0,1,1,1). The total number of configurations is

$$C = \left(4 \times \frac{3!}{3!0!0!0!}\right) + \left(12 \times \frac{3!}{2!1!0!0!}\right) + \left(4 \times \frac{3!}{1!1!1!0!}\right)$$

$$= 4 \times 1 + 12 \times 3 + 4 \times 6 = 4 + 36 + 24 = 64.$$

To check this result, an alternative route to determining the total number of possible states is to multiply the number of possibilities for each throw: $4 \times 4 \times 4 = 64$. An implication of this analysis is that you are nine times more likely to get a matching pair of dice when throwing three four-sided dice than to match all three.

10. The number of configurations in which all the dice are different is given by

$$C = N!/(n_1!\ n_2!\ n_3!\ \ldots) = N! / (1!\ 1!\ 1!\ \ldots) = N!.$$

N	C = N!
1	1
2	2
3	6
4	24
5	120
6	720

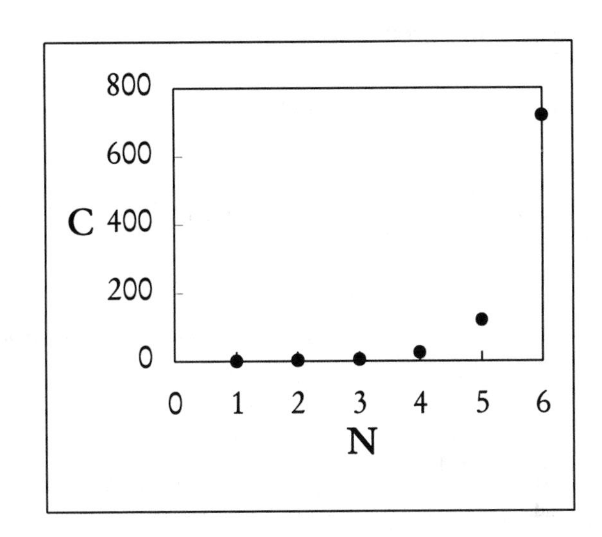

11. For half of the sample (500 g), the temperature increases from 300 to 310 K. For this process,

$$\Delta S = \left(\frac{500\ \text{g}}{18.015\ \text{g mol}^{-1}}\right) \times 8.314\ \text{J K}^{-1}\text{mol}^{-1} \times \ln(\frac{310}{300}) = 7.57\ \text{J K}^{-1}.$$

For the other 500 g, the temperature decreases from 300 to 290 K.

$$\Delta S = \left(\frac{500\ \text{g}}{18.015\ \text{g mol}^{-1}}\right) \times 8.314\ \text{J K}^{-1}\text{mol}^{-1} \times \ln(\frac{290}{300}) = -7.82\ \text{J K}^{-1}.$$

The total change in entropy is $\Delta S = 7.57 - 7.82 = -0.25\ \text{J K}^{-1}$.

Now we apply [1-9b], setting the probability of $P_1 = 1.0$ (since the water is in that state initially, its probability is unity).

$$P_2 = 1.0 \times e^{-0.25\ \text{J K}^{-1}/1.381 \times 10^{-23}\ \text{J K}^{-1}} = e^{-1.8 \times 10^{22}}.$$

Although the probability is vanishingly small, it is not identically zero. There is some (infinitesimally) small chance for the water to spontaneously separate.

12. Equation [1-9b] relates the ratio of the probabilities of two states to the difference in entropy between the two states.

$$\frac{P_2}{P_1} = 0.01 = e^{\Delta S/k}$$

We can rearrange this expression by taking the natural log of both sides to relate the given probability to the difference in entropy.

$$k \ln 0.01 = \Delta S$$

Now, we can use the relationship given in Problem 11 to relate ΔS to ΔT. Follow similar steps as in the solution to Problem 11 to arrive at

$$\Delta S_{tot} = \left(\frac{500\text{ g}}{18.015\text{ g mol}^{-1}}\right) 8.314\text{ J mol}^{-1}\text{ K}^{-1}\left[\ln\frac{300+\Delta T}{300} + \ln\frac{300-\Delta T}{300}\right]$$

$$\Delta S_{tot} = 230.8\text{ J K}^{-1}\left[\ln\left(1+\frac{\Delta T}{300}\right) + \ln\left(1-\frac{\Delta T}{300}\right)\right].$$

Replace the ln terms using the series expansion for ln(1+x):

$$\ln(1+x) = x - x^2/2 + \ldots \approx x - x^2/2$$

(when neglecting all third- and higher order terms).

$$\Delta S_{tot} \approx 230.8\text{ J K}^{-1}\left[\left(\frac{\Delta T}{300} - \frac{1}{2}\left(\frac{\Delta T}{300}\right)^2\right) + \left(\frac{-\Delta T}{300} - \frac{1}{2}\left(\frac{-\Delta T}{300}\right)^2\right)\right]$$

$$\approx -230.8\text{ J K}^{-1} \times \left(\frac{\Delta T}{300}\right)^2 = (-2.564 \times 10^{-3}\text{ J K}^{-3})\,(\Delta T)^2.$$

Finally, use the expression from above to relate the probabilities to ΔT.

$$k \ln 0.01 = \Delta S = -2.564 \times 10^{-3}\text{ J K}^{-1}(\Delta T)^2.$$

$$(\Delta T)^2 = (1.381 \times 10^{-23}\text{ J K}^{-1} \times -4.605) / -2.564 \times 10^{-3}\text{ J K}^{-3}$$

$$= 2.481 \times 10^{-20}\text{ K}^2.$$

$$\Delta T = 1.575 \times 10^{-10}\text{ K}.$$

13. Stirling's approximation is often be invoked in physical chemistry, and it is helpful to understand the accuracy of the two forms. We saw in Problem 6

that as N becomes large (greater than 1000) the relative difference between the two approximations becomes insignificant. The table below illustrates the absolute accuracy of the longer form of the approximation [1-14], even at small values of N.

N	ln N!	N ln N – N	N ln N – N + 1/2 ln(2πN)
5	4.78749	3.04719	4.77085
10	15.1044	13.0258	15.0961
15	27.8993	25.6208	27.8937
25	42.3356	39.9147	42.3314

14. For each oscillator, the energy of an oscillator in state i is given by

$$\varepsilon_i = (i + 1/2)\, h\nu \; [1\text{-}2].$$

The total energy is given by [1-12]

$$E_{tot} = \sum_i n_i \varepsilon_i = \sum_i n_i (i + 1/2)\, h\nu = (5 + \sum_i n_i \times i)\, h\nu.$$

For the state (2,1,1,5,0,1,0,0,…),

$$\sum_i n_i \times i = 2 \times 0 + 1 \times 1 + 1 \times 2 + 5 \times 3 + 0 \times 4 + 1 \times 5 + 0 \times 6 + \ldots = 23.$$

$$E_{tot} = (5 + 23)\, h\nu = 28\, h\nu.$$

Five states of 10 oscillators with the same energy are:

$$(4,0,0,1,5,0,0,0,\ldots): \sum_i n_i \times i = 4 \times 0 + 1 \times 3 + 5 \times 4 = 0 + 3 + 20 = 23.$$

$$(5,0,0,1,0,4,0,0,\ldots): \sum_i n_i \times i = 5 \times 0 + 1 \times 3 + 4 \times 5 = 0 + 3 + 20 = 23.$$

$$(3,0,0,6,0,1,0,0,\ldots): \sum_i n_i \times i = 3 \times 0 + 6 \times 3 + 1 \times 5 = 0 + 18 + 5 = 23.$$

$$(2,0,1,7,0,0,0,0,\ldots): \sum_i n_i \times i = 2 \times 0 + 1 \times 2 + 7 \times 3 = 0 + 2 + 21 = 23$$

$$(6,0,0,0,0,1,3,0,\ldots): \sum_i n_i \times i = 6 \times 0 + 1 \times 5 + 3 \times 6 = 0 + 5 + 18 = 23$$

For each, the number of the oscillators is 10 and $\sum_i n_i \times i = 23$.

For the state (1,2,3,4,0,0,0,…), there are still 10 oscillators, but since

$$\sum_i n_i \times i = 1 \times 0 + 2 \times 1 + 3 \times 2 + 4 \times 3 = 0 + 2 + 6 + 12 = 20 \neq 23,$$

thus, E_{tot} is not the same and [1-12] does not hold.

15. kT at 300 K = 0.695 cm^{-1} K^{-1} × 300 K = 208.6 cm^{-1}.

$$\frac{P_2}{P_1} = 0.5 = e^{-\Delta E(cm^{-1})/208.6\ cm^{-1}}.$$

Take the natural log of both sides and multiply by –1 to find ΔE.

$$\Delta E = -\ln 0.5 \times 208.6\ cm^{-1}$$
$$= 144.6\ cm^{-1}.$$

See Figure 1.1 for the typical spacing of energy levels. Rotational energy levels are typically 100s of cm^{-1}, whereas vibrational energy levels are typically 1000s of cm^{-1}. Therefore, rotational levels above the ground state are more likely to be populated at room temperature than vibrational states above the ground state.

16. Using the value of kT from the previous problem, we get

$$\frac{P_2}{P_1} = e^{-20{,}000\ cm^{-1}/208.6\ cm^{-1}} = e^{-95.88} = 2.296 \times 10^{-42}.$$

$$\frac{P_2}{P_1} = 0.1 = e^{-20{,}000\ cm^{-1}/kT}.$$

We find kT by taking the natural log of both sides and rearranging.

$$kT = -20{,}000\ cm^{-1} / \ln 0.1 = 8686\ cm^{-1}.$$

$$T = 8686\ cm^{-1} / k$$
$$= 8686\ cm^{-1} / 0.6950\ cm^{-1}\ K^{-1}$$
$$= 12{,}498\ K.$$

17. $E_0 = 20\ cm^{-1} \times J \times (J+1) = 20\ cm^{-1} \times 0 \times 1 = 0\ cm^{-1}$.

$g_0 = (2 \times J) + 1 = (2 \times 0) + 1 = 1$.

Using [1-11] we get

$$\frac{P_J}{P_0} = \frac{(2J+1)\times e^{-20J(J+1)\,\text{cm}^{-1}/208.3\,\text{cm}^{-1}}}{1\times e^{-0}} = (2J+1)\ e^{-20J(J+1)/208.3}.$$

J	P_J / P_0
0	1.000
1	2.477
2	2.813
3	2.215
4	1.323
5	0.620
6	0.232
7	0.070
8	0.017
9	0.003
10	0.001

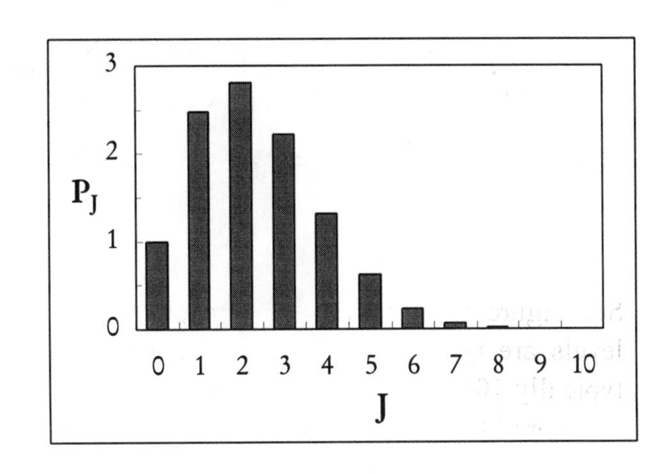

18. In only two dimensions, $\varepsilon_{\text{atom}} = \frac{1}{2m}(p_x^2 + p_y^2)$ (compare with [1-31]). Thus, the integral expression for Q [1-32] now becomes

$$Q = C\int_{-\infty}^{\infty}\int_{-\infty}^{\infty} e^{-(p_x^2+p_y^2)\beta/2m}\,dp_x\,dp_y$$

$$= C\int_{-\infty}^{\infty} e^{-\beta p_x^2/2m}\,dp_x \int_{-\infty}^{\infty} e^{-\beta p_y^2/2m}\,dp_y$$

$$= C\left(\frac{2\pi m}{\beta}\right),$$

where C is the undetermined constant of proportionality. (We evaluate these integrals by substituting $\beta/2m = c$ and applying

$$\int_{0}^{\infty} e^{-cx^2}\,dx = \frac{1}{2}\sqrt{\frac{\pi}{c}}$$

from Appendix IV). Since e^{-cx^2} is an even function, the integral from $-\infty$ to ∞ is just twice the integral from 0 to ∞).

$$E = -\frac{N}{Q}\frac{dQ}{d\beta} = -\frac{N}{Q}\frac{d}{d\beta}C\left(\frac{2\pi m}{\beta}\right) = -\frac{N}{Q}C\left(-\frac{2\pi m}{\beta^2}\right)$$

$$= -\frac{N}{Q}C\left(-\frac{2\pi m}{\beta}\right)\left(\frac{1}{\beta}\right) = -\frac{N}{Q}\left(-\frac{Q}{\beta}\right) = \frac{N}{\beta} = NkT.$$

since $\beta = 1/kT$ (see Section 1.6).

Chapter 2

Ideal and Real Gases

Exercises

1. $$P = \frac{nRT}{V} = \frac{1 \text{ mol} \times 8.2057 \times 10^{-2} \text{ L atm K}^{-1} \text{ mol}^{-1} \times 300 \text{ K}}{1 \text{ L}} = 24.617 \text{ atm.}$$

$$24.617 \text{ atm} \times \frac{101{,}325 \text{ Pa}}{1 \text{ atm}} = 2.494 \times 10^{6} \text{ Pa.}$$

2. $$P = \frac{nRT}{V} = \frac{3 \text{ mol} \times 8.206 \times 10^{-2} \text{ L atm K}^{-1} \text{ mol}^{-1} \times 100 \text{ K}}{100 \text{ L}} = 0.24617 \text{ atm.}$$

$$0.2461 \text{ atm} \times \frac{101{,}325 \text{ Pa}}{1 \text{ atm}} = 2.494 \times 10^{4} \text{ Pa.}$$

3.

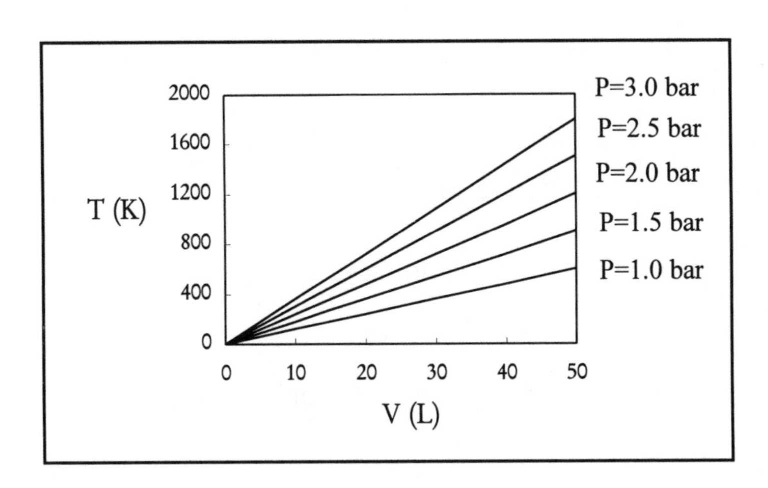

4. To solve this problem we must apply the ideal gas law, PV = nRT. Since the volume and the number of moles of gas are assumed to be constant, the initial and final conditions reduce to $P_i / T_i = P_f / T_f$. However, to apply this relationship, we must convert the given data to absolute pressures and temperatures. The measured pressure of the tire is the *gauge pressure*; the gauge pressure is the difference between the total pressure in the tire and atmospheric pressure. Therefore, the total pressure inside the tire initially is the sum of the gauge pressure and atmospheric pressure.

   $P_{initial}$ = 40 lb in.$^{-2}$ + 14.7 lb in.$^{-2}$ (atmospheric pressure) = 54.7 lb in.$^{-2}$.

   (Atmospheric pressure in lb in^{-2} (psi) can be found from Table VI.1 in Appendix VI.)

   Next, we must convert the temperatures from °F to K.

   $$(70\ °F - 32\ °F) \times (5/9) = 21.1\ °C;\ 21.1\ °C + 273.2 = 294.3\ K.$$
   $$(105\ °F - 32\ °F) \times (5/9) = 40.6\ °C;\ 40.6\ °C + 273.2 = 313.8\ K.$$

   P_{final} = 54.7 lb in.$^{-2}$ × (313.8 K / 294.3 K) = 58.3 lb in.$^{-2}$.

   P_{final} (gauge pressure) = 58.3 lb in.$^{-2}$ – 14.7 lb in.$^{-2}$ = 43.6 lb in.$^{-2}$.

5. The mass of an O_2 molecule is given by

   $$2 \times 15.9949\ \text{amu} \times 1.6606 \times 10^{-27}\ \text{kg amu}^{-1} = 5.3122 \times 10^{-26}\ \text{kg}.$$

   We find the mean speed, ⟨s⟩, using [2-17]:

   $$\langle s \rangle = \sqrt{\frac{8kT}{\pi m}} = \sqrt{\frac{(8)(1.381 \times 10^{-23}\ \text{J K}^{-1})(300\ \text{K})}{(3.14159)\ (5.3122 \times 10^{-26}\ \text{kg})}} = 445.6\ \text{m s}^{-1}.$$

   $$\langle s \rangle = \sqrt{\frac{(8)(1.381 \times 10^{-23}\ \text{J K}^{-1})(400\ \text{K})}{(3.14159)(5.3122 \times 10^{-26}\ \text{kg})}} = 514.5\ \text{m s}^{-1}.$$

   The most probable speed is given in Example 2.1 [p. 42].

$$s_{max} = \sqrt{\frac{2kT}{m}} = \sqrt{\frac{(2)\,(1.381 \times 10^{-23}\ \mathrm{J\,K^{-1}})\,(300\ \mathrm{K})}{5.3122 \times 10^{-26}\ \mathrm{kg}}} = 394.9\ \mathrm{m\,s^{-1}}.$$

$$s_{max} = \sqrt{\frac{(2)\,(1.381 \times 10^{-23}\ \mathrm{J\,K^{-1}})\,(400\ \mathrm{K})}{5.3122 \times 10^{-26}\ \mathrm{kg}}} = 456.0\ \mathrm{m\,s^{-1}}.$$

The root-mean-squared speed is given by [1-42]:

$$\langle s^2 \rangle^{1/2} = \sqrt{\frac{3kT}{m}} = \sqrt{\frac{(3)\,(1.381 \times 10^{-23}\ \mathrm{J\,K^{-1}})\,(300\ \mathrm{K})}{5.3122 \times 10^{-26}\ \mathrm{kg}}} = 483.7\ \mathrm{m\,s^{-1}}.$$

$$\langle s^2 \rangle^{1/2} = \sqrt{\frac{(3)\,(1.381 \times 10^{-23}\ \mathrm{J\,K^{-1}})\,(400\ \mathrm{K})}{5.3122 \times 10^{-26}\ \mathrm{kg}}} = 558.5\ \mathrm{m\,s^{-1}}.$$

Note that just as in Figure 2.5 [page 41], $s_{max} < \langle s \rangle < \langle s^2 \rangle^{1/2}$.

6. Ideal gas: $T = \frac{PV}{nR} = \frac{(10^5\ \mathrm{Pa})\,(9.8692 \times 10^{-6}\ \frac{\mathrm{atm}}{\mathrm{Pa}})\,(20.0\ \mathrm{L})}{(1\ \mathrm{mol})\,(8.2057 \times 10^{-2}\ \frac{\mathrm{L\,atm}}{\mathrm{K\,mol}})} = 240.543\ \mathrm{K}$

   van der Waals gas: $T = \left(P + \frac{an^2}{V^2}\right) \times \frac{V - nb}{nR}.$

   Note that in Table 2.2, the units of the a parameters are $\mathrm{L^2\ bar\ mol^{-2}}$. To simplify, we convert the given pressure of 10^5 Pa to 1 bar (Table VI.1), use this value in the van der Waals equation, then convert to atmospheres.

   Argon:

$$T = \left(1\ \mathrm{bar} + (1.355\ \frac{\mathrm{L^2\,bar}}{\mathrm{mol^2}})(1\ \mathrm{mol})^2 / (20\ \mathrm{L})^2\right) \times 0.98692\,\frac{\mathrm{atm}}{\mathrm{bar}}$$

$$\times \left(20\ \mathrm{L} - (1\ \mathrm{mol})(0.03201\,\frac{\mathrm{L}}{\mathrm{mol}})\right) / (1\ \mathrm{mol})(8.205 \times 10^{-2}\,\frac{\mathrm{L\,atm}}{\mathrm{K\,mol}}) = 240.972\ \mathrm{K}$$

Methane:

$$T=\left(1\text{ bar}+\left(2.300\,\frac{L^2\text{bar}}{\text{mol}^2}\right)\left(1\text{ mol}^2\right)/\,(20\text{ L})^2\right)\times 0.98692\,\frac{\text{atm}}{\text{bar}}$$

$$\times\left(20\text{ L}-(1\text{ mol})\left(0.04301\,\frac{L}{\text{mol}}\right)\right)/\,(1\text{ mol})\left(8.2057\times10^{-2}\,\frac{\text{Latm}}{\text{Kmol}}\right)=241.406\text{ K}$$

Acetylene:

$$T=\left(1\text{ bar}+(4.516\,\frac{L^2\text{ bar}}{\text{mol}^2})(1\text{ mol})^2/\,(20\text{ L})^2\right)\times 0.98692\frac{\text{atm}}{\text{bar}}$$

$$\times\left(20\text{ L}-(1\text{ mol})(0.05220\frac{L}{\text{mol}})\right)/\,(1\text{ mol})(8.2057\times10^{-2}\,\frac{\text{L atm}}{\text{K mol}})=242.624\text{ K}$$

7. By rearranging [2-31], we express pressure in terms of n, V, T:

$$P=\frac{nRT}{V-nb}-\frac{an^2}{V^2}.$$

$$P=\frac{(4\text{ mol})(8.314\times10^{-2}\,\frac{\text{L bar}}{\text{K mol}})(400\text{ K})}{[100\text{ L}-(4\text{ mol})(0.09044\text{ L mol}^{-1})]}-\frac{(9.385\frac{L^2\text{ bar}}{\text{mol}^2})(4\text{ mol})^2}{(100\text{ L})^2}$$

$= 1.3201$ bar.

Additional Exercises

8. The fraction *f* of molecules with speeds greater than s_{max}, the most probable speed, is equal to the total probability that a molecule's speed is greater than s_{max}. The probability that a molecule has a speed between s and s + ds, θ and θ + dθ, and ϕ and ϕ + dϕ is given by the differential probability [2-16]:

$$\left(\frac{m}{2\pi kT}\right)^{3/2} e^{-ms^2/2kT}\, s^2\, ds\, \sin\theta\, d\theta\, d\phi\,.$$

To calculate the total probability that a molecule's speed is greater than s_{max}, we simply integrate this differential probability over all orientations and from s_{max} to ∞:

$$f = \int_{s_{max}}^{\infty}\int_0^{\pi}\int_0^{2\pi}\left(\frac{m}{2\pi kT}\right)^{3/2} e^{-ms^2/2kT} s^2 ds \sin\theta\, d\theta\, d\phi$$

$$= \int_{s_{max}}^{\infty}\left(\frac{m}{2\pi kT}\right)^{3/2} e^{-ms^2/2kT} s^2 ds \int_0^{\pi}\sin\theta\, d\theta \int_0^{2\pi} d\phi$$

$$= 4\pi \int_{s_{max}}^{\infty}\left(\frac{m}{2\pi kT}\right)^{3/2} e^{-ms^2/2kT} s^2 ds\ .$$

This last step was accomplished by carrying out the integration over θ and ϕ. Now, we must substitute the actual value of s_{max} as the lower limit of the integration over s.

$$f = 4\pi \int_{\sqrt{2kT/m}}^{\infty}\left(\frac{m}{2\pi kT}\right)^{3/2} e^{-m\,s^2/2kT} s^2 ds\ .$$

We can carry out this integration by substituting a new variable, y, so that y $= \sqrt{m/2kT}\,s$, and therefore $dy = \sqrt{m/2kT}\,ds$. If we rewrite the previous expression by splitting up the factor of $(m/2\pi kT)^{3/2}$, we see that the substitution leads to

$$f = \frac{4\pi}{\pi^{3/2}} \int_{\sqrt{2kT/m}}^{\infty} e^{-m\,s^2/2kT}\left(\frac{m}{2kT}\right) s^2 \left(\frac{m}{2kT}\right)^{1/2} ds = \frac{4}{\sqrt{\pi}}\int_1^{\infty} e^{-y^2} y^2 dy.$$

Since this integral goes from 1 to ∞, it is not in a form which can be readily looked up, however by applying integration by parts we can arrive at a form which can be looked up

$$\int u\, dv = uv - \int v\, du.$$

Let $-y/2 = u$ and $-2y\, e^{-y^2} dy = dv$; this gives us $v = e^{-y^2}$ and $du = -1/2\, dy$.

$$f = \frac{4}{\sqrt{\pi}}\left(\left[\frac{-y\, e^{-y^2}}{2}\right]_1^{\infty} - \int_1^{\infty} e^{-y^2}\left(\frac{-dy}{2}\right)\right) = \frac{2}{\sqrt{\pi}}\left[\left(0-(-e^{-1})\right) + \int_1^{\infty} e^{-y^2}\, dy\right]$$

$$= \frac{2}{e\sqrt{\pi}} + \frac{2}{\sqrt{\pi}} \int_1^{\infty} e^{-y^2} dy.$$

(Note that in evaluating the first term at $y = \infty$, the expression reduces to $\infty \times e^{-\infty^2}$; this equals zero because e^{-y^2} goes to 0 must faster that y goes to ∞.)

The remaining integral in the second term is closely related to a special function called the error function, which is defined as

$$\text{erf}(z) = \frac{2}{\sqrt{\pi}} \int_0^{z} e^{-y^2} dy = 1 - \frac{2}{\sqrt{\pi}} \int_z^{\infty} e^{-y^2} dy.$$

$$\frac{2}{\sqrt{\pi}} \int_z^{\infty} e^{-y^2} dy = 1 - \text{erf}(z)$$

The value of erf(1) can be looked up in a mathematical handbook, so that finally,

$$f = \frac{2}{e\sqrt{\pi}} + 1 - \text{erf}(1) = 0.4151 + 1 - 0.8427 = 0.5724.$$

This result says that just more than half of all the molecules in an ideal gas have a speed greater than the most probable speed. If you examine Figure 2.5, this agrees with the diagram, which shows the distribution of speed to be skewed in favor of the higher velocities (which is why $\langle s \rangle > s_{max}$).

9. The first step is to determine the mean kinetic energy. Since kinetic energy $= 1/2\ ms^2$, the mean kinetic energy is just

$$\langle 1/2\ ms^2 \rangle = 1/2\ m\ \langle s^2 \rangle = 1/2\ m \times (3kT/m) = 3/2\ kT,$$

where we find the value of $\langle s^2 \rangle$ from [1-42]. Twice the mean kinetic energy is $2 \times 3/2\ kT = 3kT$, and by equating that with $1/2\ ms^2$ we see that the speed of a molecule with twice the mean kinetic energy is $(6kT/m)^{1/2}$.

Now, to answer the problem, we restate it as follows: What fraction of molecules has a speed greater than $(6kT/m)^{1/2}$? Following the same steps of the previous problem, we arrive at the expression

$$f = 4\pi \int_{\sqrt{6kT/m}}^{\infty} \left(\frac{m}{2\pi kT}\right)^{3/2} e^{-m s^2/2kT}\ s^2\ ds\ .$$

Making the same substitution as in Problem 8, we get

$$f = \frac{4\pi}{\pi^{3/2}} \int_{\sqrt{6kT/m}}^{\infty} e^{-m s^2/2kT} \left(\frac{m}{2kT}\right) s^2 \left(\frac{m}{2kT}\right)^{1/2} ds = \frac{4}{\sqrt{\pi}} \int_{\sqrt{3}}^{\infty} e^{-y^2} y^2\, dy$$

$$= \frac{4}{\sqrt{\pi}} \left(\left[\frac{-y\, e^{-y^2}}{2} \right]_{\sqrt{3}}^{\infty} - \int_{\sqrt{3}}^{\infty} e^{-y^2} \left(\frac{-dy}{2} \right) \right)$$

$$= \frac{2}{\sqrt{\pi}} \left[\left(0 - (-\sqrt{3}e^{-3}) \right) + \int_{\sqrt{3}}^{\infty} e^{-y^2}\, dy \right]$$

$$f = \frac{2}{e^3}\sqrt{\frac{3}{\pi}} + 1 - \text{erf}(\sqrt{3}) = 0.0973 + 1 - 0.9856 = 0.1117.$$

This result indicates that most of the molecules in an ideal gas have energies less than twice the mean kinetic energy (since only 11.17 % have kinetic energies greater than twice the mean kinetic energy).

10. $v_x^2 = s^2 \sin^2\theta \cos^2\phi$.

$v_y^2 = s^2 \sin^2\theta \sin^2\phi$.

$v_x^2 + v_y^2 = s^2 \sin^2\theta (\cos^2\phi + \sin^2\phi) = s^2 \sin^2\theta$ [since $\cos^2\phi + \sin^2\phi = 1$].

$v_z^2 = s^2 \cos^2\theta$.

$(v_x^2 + v_y^2) + v_z^2 = s^2 (\cos^2\theta + \sin^2\theta) = s^2$ [since $\cos^2\theta + \sin^2\theta = 1$].

$\therefore v_x^2 + v_y^2 + v_z^2 = s^2$.

11. Following [2-17]

$$\langle s^4 \rangle = 4\pi \left(\frac{m}{2\pi kT} \right)^{3/2} \int_0^{\infty} s^6\, e^{-m s^2/2kT}\, ds.$$

This integral is of the type given in Appendix IV, Section 3, if we let c = m/2kT and n = 3. Substituting these values into the formula, we get

$$\langle s^4 \rangle = 4\pi\left(\frac{m}{2\pi kT}\right)^{3/2} \frac{5\times 3\times 1}{2^4}\sqrt{\pi}\left(\frac{2kT}{m}\right)^{7/2} = 15\left(\frac{kT}{m}\right)^2.$$

12. First, we rewrite the equation of state so that pressure is in terms of V,T,n:

$$P = \frac{nRT}{V-a}.$$

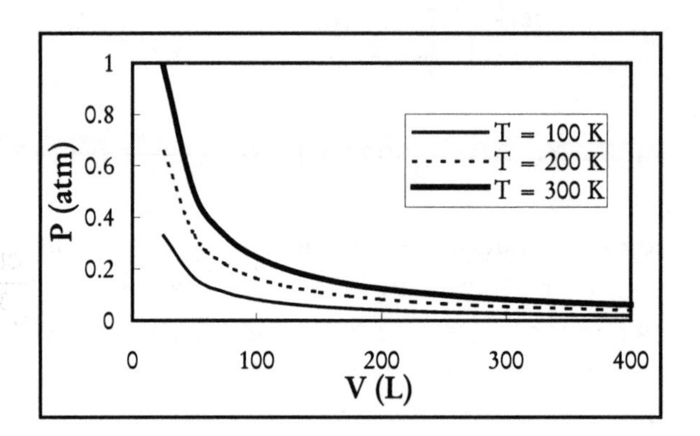

To compare with ideal gas isotherms, we rewrite V in terms of P,T, n:

$$V = \frac{nRT}{P} + a.$$

When compared to ideal gas isotherms (where V = nRT/P), it can be seen that $V_{real} = V_{ideal} + a$. Therefore, the isotherms of this equation of state will be identical with those of the ideal gas equation of state, merely displaced to the right by a.

13. From Table 2.2, a = 5.537 L^2 bar mol^{-2}, and b = 0.03049 L mol^{-1}. At T = 380 K , the van der Waals pressure–volume isotherm is defined by

$$P(\text{atm}) = \frac{(1\text{ mol})(8.314\times 10^{-2}\ \frac{\text{L bar}}{\text{K mol}})(380\text{ K})}{(\text{V (L)} - (1\text{ mol})(0.03049\text{ L mol}^{-1}))} - \frac{(5.537\,\frac{\text{L}^2\text{ bar}}{\text{mol}^2})(1\text{ mol})^2}{(\text{V (L)})^2}.$$

In the figure below, note that at small volumes (high density), the van der Waals equation predicts negative pressures, which are physically impossible. At high density, the van der Waals equation becomes inadequate to describe the behavior of condensed phases (liquids).

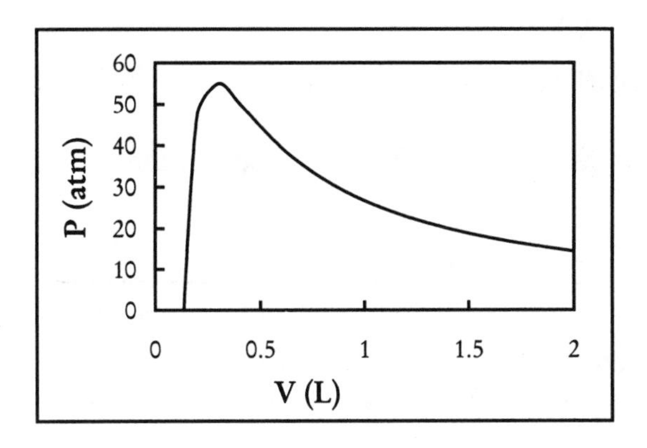

14. $Z = \dfrac{PV}{nRT} = 1 + \dfrac{b}{V} + \dfrac{c}{V^2} + \dfrac{d}{V^3} = 1 - \dfrac{100\ cm^3}{V\ (cm^3)} + \dfrac{7200\ cm^6}{(V\ (cm^3))^2}$

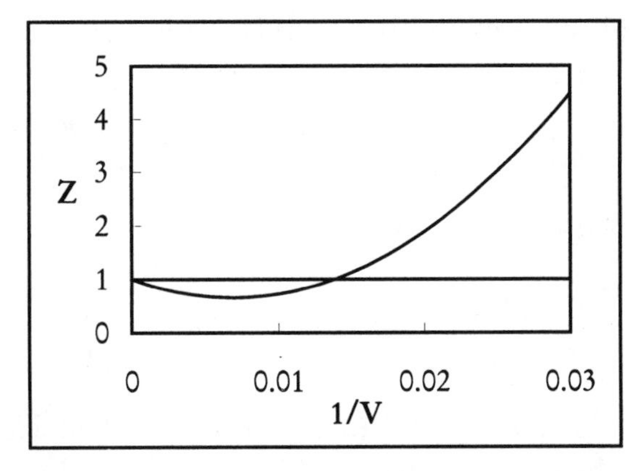

See how $Z \rightarrow 1$ as $1/V \rightarrow 0$, which is when $V \rightarrow \infty$. When $V \rightarrow \infty$, all gases behave ideally, since on average the individual gas molecules are so far from one another they have no interactions. As the volume is decreased ($1/V$ increases), the compressibility factor decreases ($Z < 1$). When V is large, $1/V^2 \ll 1/V$ and the b/V term will dominate; therefore, this decrease in Z indicates that $b < 0$. When $Z < 1$, then $PV < nRT$, which can be true if $V_{real} < V_{ideal}$. V_{real} will be less than V_{ideal} if the forces between molecules are attractive and hold the molecules together more closely. Thus, $b < 0$ indicates that the first interactions to become important (in other words, that have the longest range) are attractive, not repulsive.

15. The constant-pressure coefficient of thermal expansion [2-35] is defined to be

$$\alpha = \frac{1}{V}\left(\frac{\partial V}{\partial T}\right)_P.$$

The Bertholet gas equation [2-30] can be written as

$$V = \frac{nRT}{P} + a - \frac{ab}{T^2}$$

(It is necessary to rewrite it in this form so that the derivative of V with respect to T can be taken; this requires that V be written as a function of T.)

$$\alpha = \frac{1}{V}\left[\frac{\partial}{\partial T}\left(\frac{nRT}{P} + a - \frac{ab}{T^2}\right)\right]_P = \frac{1}{V}\left(\frac{nR}{P} + \frac{2ab}{T^3}\right) = \frac{nR}{PV} + \frac{2ab}{VT^3}.$$

16. The isothermal compressibility [2-36] is defined as

$$\beta = -\frac{1}{V}\left(\frac{\partial V}{\partial P}\right)_T.$$

For an ideal gas, where $V = nRT/P$,

$$\beta = -\frac{1}{V}\left[\frac{\partial}{\partial P}\left(\frac{nRT}{P}\right)\right]_T = -\frac{1}{V}\left(-\frac{nRT}{P^2}\right) = -\frac{P}{nRT}\left(-\frac{nRT}{P^2}\right) = \frac{1}{P}$$

For the virial equation [2-29] (see note in Problem 15 for why the virial equation must be written in this form),

$$\beta = -\frac{1}{V}\left[\frac{\partial}{\partial P}\left(\frac{nRT}{P} + a + bP + cP^2\right)\right]_T$$

$$= -\frac{1}{V}\left(-\frac{nRT}{P^2} + b + 2cP\right)$$

$$= -\frac{1}{PV}\left(-\frac{nRT}{P} + bP + 2cP^2\right)$$

$$= -\frac{1}{PV}\left(-V + a + bP + cP^2 + bP + 2cP^2\right)$$

$$= -\frac{1}{PV}\left(-V + a + 2bP + 3cP^2\right)$$
$$= \frac{1}{P} - \frac{a}{PV} - \frac{2b}{V} - \frac{3cP}{V}$$

17.
$$\frac{\alpha}{\beta} = \frac{\frac{1}{V}\left(\frac{\partial V}{\partial T}\right)_P}{-\frac{1}{V}\left(\frac{\partial V}{\partial P}\right)_T} = -\left(\frac{\partial V}{\partial T}\right)_P \Big/ \left(\frac{\partial V}{\partial P}\right)_T$$

Now, we apply relationship [I-15] from Appendix I, substituting V for z, T for x, and P for y:

$$\frac{\alpha}{\beta} = -\left(\frac{\partial V}{\partial T}\right)_P \Big/ \left(\frac{\partial V}{\partial P}\right)_T = \left(\frac{\partial P}{\partial T}\right)_V$$

You can now apply this to the van der Waals equation.

$$\frac{\alpha}{\beta} = \left(\frac{\partial P}{\partial T}\right)_V = \left(\frac{\partial}{\partial T}\left(\frac{nRT}{V - nb} - \frac{an^2}{V^2}\right)\right)_V = \frac{nR}{V - nb}$$

18. We follow the example of Appendix I (after [I-3]), to generate a series expansion for e^x:

$$e^x = e^0 + x\left.\frac{de^x}{dx}\right|_{x=0} + \frac{x^2}{2}\left.\frac{d^2e^x}{dx^2}\right|_{x=0} + \frac{x^3}{6}\left.\frac{d^3e^x}{dx^3}\right|_{x=0} + \cdots$$

$$e^x = 1 + x + \frac{x^2}{2} + \frac{x^3}{6} + \cdots \quad \left(\text{since } \frac{de^x}{dx} = e^x, \text{ and } e^0 = 1\right).$$

The Dieterici equation is $P(V - b) = nRTe^{-an/VRT}$. Using the series expansion for e^x, we get

$$e^{-an/VRT} = 1 - \frac{an}{VRT} + \frac{(an)^2}{2(VRT)^2} + \cdots.$$

If we keep only the first term in the series, we approximate $e^{-an/VRT} \approx 1$, and then the Dieterici equation reduces to $P(V - b) = nRT$, which can be rearranged to

$$P = \frac{nRT}{V - b}.$$

This is similar to the van der Waals equation with the parameter a set to 0.

If we keep two terms, the equation becomes

$$P(V - b) = nRT\left(1 - \frac{an}{VRT}\right) \Rightarrow P = \frac{nRT}{V - b} - \frac{an^2}{V(V - b)}.$$

This now very closely resembles the full van der Waals equation.

19. There are eight thermodynamic state functions, each of which can be written in terms of any other two functions. For each of the eight functions, there are seven possibilities to choose from for the first function on which it can depend, and this leaves six possibilities for the second function on which it can depend. However, these choices will count each possibility twice. As an example, let's choose to express T in terms of the other functions. If we initially choose P as the first function, the possibilities are T(P,V), T(P,S), T(P,U), ... But choosing U as the first function leads to T(U,P), T(U,V), ... Now, T(P,U) and T(U,P) are identical. Therefore, the total number of unique differential expressions is $8 \times 7 \times 6 / 2 = 168$.

20. We start with the relationship

$$dH = T\,dS + V\,dP \qquad [2\text{-}51].$$

But if we simply write H as a function of (S,P) by [2-40],

$$dH = \left(\frac{\partial H}{\partial S}\right)_P dS + \left(\frac{\partial H}{\partial P}\right)_S dP$$

must also be true, see [2-52]. By comparing these two expressions we see that

$$\left(\frac{\partial H}{\partial P}\right)_S = V.$$

Similarly, in Example 2.3 we see that since

$$dG = -S\,dT + V\,dP$$

and

$$dG = \left(\frac{\partial G}{\partial T}\right)_P dT + \left(\frac{\partial G}{\partial P}\right)_T dP,$$

then

$$\left(\frac{\partial G}{\partial P}\right)_T = V.$$

21. $\left(\frac{\partial U}{\partial V}\right)_S = -P \quad [2\text{-}48].$

$$\left(\frac{\partial}{\partial P}\left(\frac{\partial U}{\partial V}\right)_S\right)_S = \left(\frac{\partial}{\partial P}(-P)\right)_S = -1.$$

22. Although we have not encountered any explicit relationship between entropy and the other thermodynamic functions for a van der Waals gas, we can use one of the Maxwell relationships to express the temperature derivative of entropy with an easier derivative. Specifically, we use the relationship given in [2-56]:

$$\left(\frac{\partial S}{\partial V}\right)_T = \left(\frac{\partial P}{\partial T}\right)_V = \frac{\partial}{\partial T}\left(\frac{nRT}{V - nb} - \frac{an^2}{V^2}\right)_V = \frac{nR}{V - nb}.$$

Chapter 3

Changes of State

Exercises

1. $P = \frac{F}{A} = \frac{mg}{A} = \frac{1.0 \text{ kg} \times 9.80665 \text{ m s}^{-2}}{0.2 \text{ m}^2} = 49.033 \text{ N m}^2 = 49.033 \text{ Pa.}$

2. The initial condition of the gas is n = 1.0 mol, P = 1.0 bar, and T = 300.0 K. Since it is ideal, we can calculate the initial volume by [2-7]:

   $$V = (1.0 \text{ mol})\,(8.314 \times 10^{-2} \text{ L bar K}^{-1} \text{mol}^{-1})\,(300.0 \text{ K}) = 24.94 \text{ L.}$$

   The change in volume is simply the product of Δh and the cross–sectional area.

   $$\Delta V = (-2.0 \text{ cm})\,(25.0 \text{ cm}^2)\,(1 \text{ L} / 1000 \text{ cm}^3) = -0.05 \text{ L.}$$

   Now we can determine the final pressure:

   $$P_f = (1.0 \text{ mol})(8.314 \times 10^{-2} \text{ L bar K}^{-1} \text{mol}^{-1})(300.0 \text{ K}) / (24.94 - 0.05 \text{ L})$$
   $$= 1.002 \text{ bar.}$$

   Now that we have the final pressure, we can determine the mass required to exert the additional pressure on the gas (= $P_f - P_i$). Mass and pressure are related by $P = F/A = mg/A$; $\Delta m = (\Delta P)A/g$.

$$\Delta m = (0.002 \text{ bar})\left(\frac{10^5 \text{ Pa}}{\text{bar}}\right)\left(\frac{\text{kg m}^{-1}\text{s}^{-1}}{\text{Pa}}\right)(25 \text{ cm}^2)\left(\frac{10^{-4} \text{ m}^2}{\text{cm}^2}\right) / 9.807 \text{ m s}^{-1}$$

$$= 0.51 \text{ kg}.$$

3. We use [3-2] to calculate the work done by the reversible expansion of an ideal gas. For the first step,

$$w = -nRT \ln(V_2/V_1) = -nRT \ln(P_1/P_2).$$

($P_1/P_2 = V_2/V_1$, since n and T are held constant.)

$$w = -(1 \text{ mol})(8.314 \text{ J K}^{-1} \text{ mol}^{-1})(400 \text{ K}) \ln(1.0/0.5) = -2305 \text{ J} = -2.305 \text{ kJ}.$$

Because the ratio of the final pressure to the initial pressure is the same for every step [(0.5/1.0) = (0.25/0.5) = (0.125/0.25) = 0.5], the same amount of work is done at each step.

4. First, we rearrange the equation of state to express b in terms of n, P, T, V.

$$b = \left(\frac{PV}{n} - RT\right)\left(\frac{1}{P}\right) = \frac{V}{n} - \frac{RT}{P} = \frac{22.0 \text{ L}}{1 \text{ mol}} - \frac{(8.206 \times 10^{-2} \frac{\text{L atm}}{\text{K mol}})(300 \text{ K})}{(1.0 \text{ bar})(0.98692 \text{ atm bar}^{-1})}$$

$$= -2.9443 \text{ L mol}^{-1}$$

Now, with b determined, we can calculate the volume of the gas.

$$V = \frac{1.0 \text{ mol}}{0.5 \text{ bar}}[(8.206 \times 10^{-2} \frac{\text{L atm}}{\text{K mol}})(1.01325 \frac{\text{bar}}{\text{atm}})(300 \text{ K}) + (-2.9443 \frac{\text{L}}{\text{mol}})(0.5 \text{ bar})]$$

$$= 46.944 \text{ L}$$

5. We can calculate the work done by the gas directly from

$$w = -P_{ext}(V_2 - V_1), \qquad \text{(i)}$$

or, since this process is adiabatic [q = 0],

$$w = \Delta U = \frac{3nR}{2}(T_2 - T_1).$$

We are given enough information to specify the initial state of the ideal gas: n = 1.0 mol, T = 400 K, and P = 1 bar. We can find the initial volume with the ideal gas law,

$$V_1 = \frac{(1\ \text{mol})\ (8.206 \times 10^{-2}\ \frac{\text{L atm}}{\text{K mol}})\ (400\ \text{K})}{(1\ \text{bar})\ (0.98692\ \text{atm bar}^{-1})} = 33.26\ \text{L}.$$

But since we are given only the final pressure, 0.5 bar, the final temperature and final volume are unknown, and it appears that more information is needed. However, since this is an ideal gas, the final temperature and volume are connected by the ideal gas law.

$$P_2V_2 = nRT_2.$$

We can use this additional information in the expression above which equates work and the change in the internal energy. The final temperature can be written in terms of the final volume; then the final volume and work can be determined.

$$\Delta U = \frac{3nR}{2}(T_2 - T_1) = 3/2\ (nRT_2 - nRT_1) = 3/2\ (P_2V_2 - P_1V_1) \quad \text{(ii)}$$

(by applying the ideal gas law). Now, equate the two expressions for the work done, (i) and (ii):

$$-P_{ext}(V_2 - V_1) = 3/2(P_2V_2 - P_1V_1).$$

Since $P_2 = P_{ext}$,

$$-P_{ext}(V_2 - V_1) = 3/2(P_{ext}V_2 - P_1V_1).$$

$$(3/2\ P_1 + P_{ext})V_i = 5/2\ P_{ext}\ V_2.$$

$$V_2 = \frac{[(3/2)(1.0\ \text{bar}) + 0.1\ \text{bar}](33.26\ \text{L})}{(5/2)(0.1\ \text{bar})} = 212.9\ \text{L}.$$

$$w = -P_{ext}\ \Delta V = -(0.1\ \text{bar})(212.9\ \text{L} - 33.26\ \text{L}) = -17.964\ \text{L bar}.$$

$$-17.964 \text{ L bar} \times 100 \text{ J L}^{-1} \text{ bar}^{-1} = -1796.4 \text{ J} = -1.7964 \text{ kJ}.$$

(The conversion from L bar to J comes from

$$1 \text{ L bar}\left(\frac{10^5 \text{Pa}}{\text{bar}}\right)\left(\frac{\text{N m}^{-2}}{\text{Pa}}\right)\left(\frac{\text{dm}^3}{\text{L}}\right)\left(\frac{\text{m}^3}{1000 \text{ dm}^3}\right)\left(\frac{1 \text{ J}}{1 \text{ Nm}}\right) = 100 \text{ J}.)$$

6. $\Delta T = (2/3)(-1796.4 \text{ J}) / (8.314 \text{ J K}^{-1} \text{ mol}^{-1}) (1 \text{ mol}) = -144.0 \text{ K}.$
   $T_2 = 400 - 144.0 = 256.0 \text{ K}.$

   It is necessary to use the value of R which is in units of J, since the rest of the equation includes terms which are in units of J.

7. Because the expansion is isothermal, and the gas is ideal, the internal energy cannot change ($\Delta U = 0$). Therefore, $q + w = 0$, or $w = -q = -10\text{kJ}$. Equation [3-2] relates the work done by an ideal gas in an isothermal expansion. Since we know w, we can calculate V_2/V_1:

$$\frac{V_2}{V_1} = \exp\left(\frac{-w}{nRT}\right) = \exp\left(\frac{10{,}000 \text{ J}}{(1 \text{ mol})\left(8.3145 \text{ J K}^{-1} \text{ mol}^{-1}\right)(400 \text{ K})}\right) = 20.222.$$

8. First, we must realize that these two states cannot be at the same temperature; otherwise the product of PV would remain constant. The temperature of the first state is

$$T = (1 \text{ bar})(20 \text{ L})/(1 \text{ mol})(0.08314 \text{ L bar/K mol}) = 240.6 \text{ K}.$$

   (R in terms of L–bar can be found using the conversion factor derived in the solution to Problem 5, that 100 J = 1 L bar. Thus,

$$(8.314 \text{ J K}^{-1} \text{ mol}^{-1})(1 \text{ L–bar} / 100 \text{ J}) = 8.314 \times 10^{-2} \text{ L–bar K}^{-1} \text{ mol}^{-1}.)$$

   The temperature of the second state is

$$T = (10 \text{ bar})(1 \text{ L})/(1 \text{ mol})(0.08314 \text{ L bar/K mol}) = 120.3 \text{ K},$$

so any change of state must take the change of temperature into consideration.

An irreversible process may take any path that ensures that the final state is attained. We could begin with a piston immersed in a temperature bath at 240.6 K with sufficient weights on the piston so that a pressure of 1 bar was exerted. Then, we could add additional weight to the piston, so that a total pressure of 10 bar was exerted on the piston. Then we could lower the temperature of the water in the in any manner desired to bring the final temperature to 120.3 K.

A reversible process must meet the additional constraint that any changes are infinitesimally small, or nearly so, so that the gas in the piston always remains in equilibrium. Consider the following two-step process. As before, a piston is immersed in a temperature bath at 240.6 K, and the piston weighted so that it exerts a pressure of 1 bar on the gas, only this time the weight is applied by a pile of sand. Now, we remove the piston from the bath and completely insulate it, so that heat cannot be lost and carry out an adiabatic expansion. To accomplish this, remove the sand a grain at a time, until the expansion has cooled the gas to 120.3 K. Then we immersed the piston in the temperature bath, remove the insulation, and now carry out compression, by adding one grain of sand at a time, until a final pressure of 10 bar is achieved.

9. Recall that in Problems 5 and 6 we examined 1.0 mol of gas, whose initial state was $T_1 = 400$ K, $V_1 = 33.26$ L, and $P_1 = 1$ bar, and whose final state was $T_3 = 256.0$ K, $V_3 = 212.9$ L, and $P_3 = 0.1$ bar. The reversible process described in the previous problem can be employed here, in which a reversible adiabatic expansion first cools the gas to the final temperature, and then a reversible, isothermal expansion brings the system to the final state. For clarity, designate the intermediate state as 2. For the first step (taking the system from state 1 to state 2), the reversible, adiabatic expansion, $\Delta S = 0$, because $q_{rev} = 0$. Thus, we must first determine P_2 and V_2 of the gas in the intermediate state after adiabatically cooling to 256.0 K. Then, we can use [3-18] to determine ΔS for the isothermal expansion, and thus for the overall process.

For an adiabatic, reversible process, [3-46] relates the change in temperature to the change in volume:

$$C_V \ln\frac{T_2}{T_1} = -nR \ln\frac{V_2}{V_1}.$$

Now, we divide by –nR, take the exponential of both sides, and we arrive at

$$\left(\frac{T_2}{T_1}\right)^{\frac{-C_V}{nR}} = \frac{V_2}{V_1}.$$

For an ideal, monatomic gas, $C_V = 3nR/2$. After the adiabatic expansion,

$$V_2 = 33.26\ \mathrm{L}\left(\frac{256.0\ \mathrm{K}}{400\ \mathrm{K}}\right)^{-3/2} = 64.96\ \mathrm{L}.$$

Now, we can calculate ΔS for the isothermal expansion from 64.96 L to 212.9 L using [3-18]:

$$\Delta S = nR \ln\frac{V_3}{V_2} = (1\ \mathrm{mol})(8.314\ \mathrm{J\ K^{-1}\ mol^{-1}})(\ln\frac{212.9}{64.96}) = 9.869\ \mathrm{J\ K^{-1}}.$$

10. U for an ideal gas depends only on the temperature. Therefore, $\Delta U = 0$ for an isothermal process, which implies that $q = -w$. In Problem 3, we saw that for all three processes, $w = -2.305$ kJ; therefore $q = 2.305$ kJ for all three processes as well. ΔS is given by [3-20]:

$$\Delta S = q_{rev}/T = 2305\ \mathrm{J} / 400\ \mathrm{K} = 5.763\ \mathrm{J\ K^{-1}}.$$

Because T is constant, and q_{rev} is the same for each step, ΔS is the same for all three processes.

11. From the definition of the constant-pressure heat capacity [3-31], it follows that

$$dq = nC_P dT.$$

From the data in Table 3.1, we can approximate C_P by a polynomial and then we can integrate the last equation to determine the total change in heat:

$$\Delta q = \int dq = (2\text{ mol}) \int_{400}^{1000} \left[26.86 \frac{\text{J}}{\text{K mol}} + (0.006966 \frac{\text{J}}{\text{K}^2\text{mol}})(\text{T}) - (8.20 \times 10^{-7} \frac{\text{J}}{\text{K}^3\text{mol}})(\text{T})^2 \right] d\text{T}$$

$$= (2.0)\left[(26.86)(1000 - 400) + \frac{0.006966}{2}\left(1000^2 - 400^2\right) - \frac{8.20 \times 10^{-7}}{3}(1000^3 - 400^3) \right]$$

$$= 3.757 \times 10^4 \text{ J} = 37.57 \text{ kJ}.$$

Additional Exercises

12. Since we don't know any relationship between S, T, and V that can be differentiated directly, let's look for another derivative or derivatives that are equivalent to the one needed. We start with [2-50]:

$$dU = T\,dS - P\,dV.$$

Now, we "divide" by dT and impose constant volume:

$$\left(\frac{\partial U}{\partial T}\right)_V = T\left(\frac{\partial S}{\partial T}\right)_V - P\left(\frac{\partial V}{\partial T}\right)_V.$$

The last term is zero, because volume is constant, so

$$\left(\frac{\partial S}{\partial T}\right)_V = \frac{1}{T}\left(\frac{\partial U}{\partial T}\right)_V = \frac{1}{T}C_V = \frac{3nR}{2T}.$$

13. $w = -\int P\, dV = -\int_{V_i}^{V_f} \left(\frac{nRT}{V - nb} - \frac{an^2}{V^2} \right) dV$

(substituting for P with the van der Waals equation)

$$= \left[-nRT \ln(V - nb) - \frac{an^2}{V} \right]_{V_1}^{V_2}.$$

$$w_{ideal} = -(1\ \text{mol})\ (8.314\ \text{J K}^{-1}\ \text{mol}^{-1})\ (300\ \text{K})\ (\ln 40.0/20.0) = -1729\ \text{J}.$$

$$w_{argon} = -(1\ \text{mol})(8.314 \frac{\text{J}}{\text{K mol}})(300\ \text{K}) \ln \frac{40.0 - (1\ \text{mol})(0.03201)}{20.0 - (1\ \text{mol})(0.03201)}$$

$$- (1.355\ \text{L}^2\ \text{bar mol}^{-2})(1\ \text{mol})^2 \left(\frac{1}{40\ \text{L}} - \frac{1}{20\ \text{L}} \right) (100 \frac{\text{J}}{\text{L bar}})$$

$$= -1727\ \text{J mol}^{-1}.$$

14. The efficiency of an engine is the ratio of the net work done by the engine to the heat transferred from the high-temperature reservoir [see p. 83]. For the Carnot cycle, this efficiency is shown to be

$$\text{Efficiency} = 1 + \frac{w_3}{w_1} = 1 + \frac{-nRT_2 \ln \frac{V_4}{V_3}}{-nRT_1 \ln \frac{V_2}{V_1}} = 1 + \frac{T_2 \ln \frac{V_4}{V_3}}{T_1 \ln \frac{V_2}{V_1}}.$$

Equation [3-13] relates the volumes: $\frac{V_3}{V_2} = \frac{V_4}{V_1}$. By simple algebra, $\frac{V_1}{V_2} = \frac{V_4}{V_3}$. If this is true, then $\ln \frac{V_1}{V_2} = \ln \frac{V_4}{V_3}$, or $-\ln \frac{V_2}{V_1} = \ln \frac{V_4}{V_3}$. By substitution:

$$\text{Efficiency} = 1 - \frac{T_2}{T_1}.$$

15. We recall the four steps in the Carnot cycle: isothermal expansion (temperature constant, entropy increases, internal energy constant, pressure decreases), adiabatic expansion (temperature decreases, internal energy

decreases, entropy constant, pressure decreases), isothermal compression (temperature constant, entropy decreases, internal energy constant, pressure increases), and adiabatic compression (temperature increases, entropy constant, internal energy increases, pressure increase).

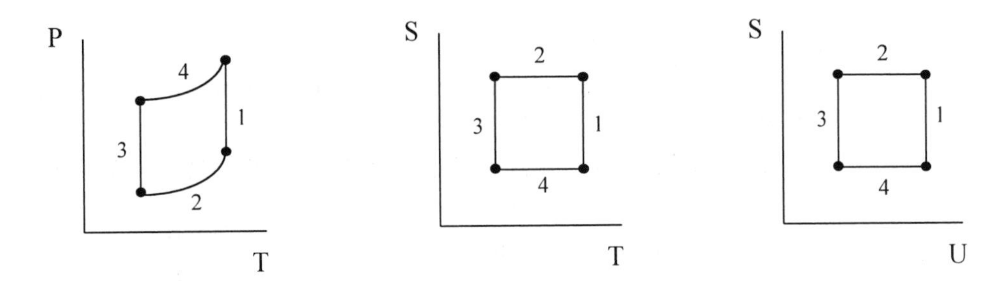

16. For this engine cycle, pressure, volume, temperature, and entropy vary as follows:
    (1) pressure constant; volume, temperature, and entropy increase
    (2) pressure and temperature decrease, volume increases, entropy constant
    (3) pressure constant; volume, temperature, and entropy decrease
    (4) pressure and temperature increase, volume decreases, entropy constant.

    To get a qualitative picture for the changes in P, V, and T which follow from the specifications of the paths, let's assume the engine uses an ideal gas. For path 1, since heat must be added to raise the temperature so that the volume increases while the pressure is constant, then the change in entropy has to be positive as well [$dS = dq_{rev}/T$]. Likewise, for path 3 the entropy must decrease. $\Delta S = 0$ for the adiabatic paths.

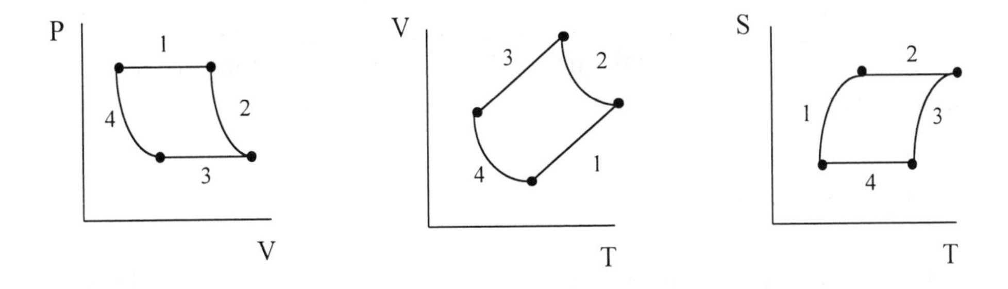

If the gas used by the engine were not ideal, the relationship between V and T for paths (1) and (3) would no longer remain linear.

17. Under conditions of constant pressure (*isobaric*),

$$dq = C_P\, dT.$$

Since $dS = dq_{rev}/T$, then

$$dS = (C_P/T)\, dT.$$

For oxygen, $C_P = (25.72 + 0.01298\ T - 3.86 \times 10^{-6}\ T^2)$ J K^{-1} mol^{-1}. To calculate ΔS, we simply insert the expression for C_P into the differential expression for dS and integrate:

$$\Delta S = \int_{300}^{400} \frac{25.72 + 0.01298\ T - 3.86 \times 10^{-6}\ T^2}{T}\, dT$$

$$= 25.72 \ln\frac{400}{300} + (0.01298)(400 - 300) - \frac{3.86 \times 10^{-6}}{2}(400^2 - 300^2)$$

$$= 8.56\ J\ K^{-1}.$$

18. We start with the expression at the top of p. 104:

$$\mu_{JT} = -\frac{1}{C_P}\left(\frac{\partial H}{\partial P}\right)_T.$$

To express the partial derivative in terms of P,V, and T only, we write the differential of dH:

$$dH = T\, dS + V\, dP.$$

Now, we "divide" by dP and impose constant temperature:

$$\left(\frac{\partial H}{\partial P}\right)_T = T\left(\frac{\partial S}{\partial P}\right)_T + V\left(\frac{\partial P}{\partial P}\right)_T = T\left(\frac{\partial S}{\partial P}\right)_T + V.$$

To eliminate the partial derivative of the entropy, we use the "thermodynamic compass" in Figure 2.8 to show that

$$\left(\frac{\partial S}{\partial P}\right)_T = -\left(\frac{\partial V}{\partial T}\right)_P.$$

(Following the directions on page 61, start at S and go counter clockwise to select P and T; since P is on the lower half, we must make this derivative the negative of itself. This derivative will be equal to the derivative specified by starting at V and selecting T and P: the derivative of V with respect to T at constant P. Thus, we arrive at the equation given above.)

Substituting this into the previous expression, we arrive at

$$\left(\frac{\partial H}{\partial P}\right)_T = -T\left(\frac{\partial V}{\partial T}\right)_P + V.$$

This lets us express the Joule-Thomson coefficient as

$$\mu_{JT} = \frac{1}{C_P}\left[T\left(\frac{\partial V}{\partial T}\right)_P - V\right]$$

This partial derivative has been evaluated for a van der Waals gas [3-43]:

$$\left(\frac{\partial V}{\partial T}\right)_P = \frac{nRV^3}{PV^3 - an^2V + 2abn^3}.$$

(To get this form of [3-43], we must multiply the numerator and denominator by V^3.)

$$\mu_{JT} = \frac{1}{C_P}\left(\frac{nRTV^3}{PV^3 - an^2V + 2abn^3} - V\right).$$

19. The internal pressure of a gas is given by [3-40]:

$$\left(\frac{\partial U}{\partial V}\right)_T = T\left(\frac{\partial P}{\partial T}\right)_V - P.$$

To determine the internal pressure of a Bertholet gas, we first take the partial derivative of P with respect to T at constant V (which requires writing the equation of state as P(V,T):

$$P = \frac{nRT^3}{(V-a)T^2 + b}.$$

$$\left(\frac{\partial P}{\partial T}\right)_V = \left(\frac{3nRT^2}{(V-a)T^2 + b}\right) - \left(\frac{nRT^3}{[(V-a)T^2 + b]^2}\right)[2(V-a)T]$$

$$= \frac{3}{T}\left(\frac{nRT^3}{(V-a)T^2 + b}\right) - \left(\frac{nRT^3}{(V-a)T^2 + b}\right)\frac{[2(V-a)T}{(V-a)T^2 + b}$$

$$= P\left(\frac{3}{T} - \frac{2(V-a)T}{(V-a)T^2 + b}\right) = P\left(\frac{(V-a)T^2 + 3b}{(V-a)T^3 + bT}\right).$$

$$\left(\frac{\partial U}{\partial V}\right)_T = T\left(\frac{\partial P}{\partial T}\right)_V - P = TP\left(\frac{(V-a)T^2 + 3b}{(V-a)T^3 + bT}\right) - P = \frac{2bP}{(V-a)T^2 + b}.$$

20. Equation [3-37] gives a definition for $C_P - C_V$.

$$C_P - C_V = \left(\frac{\partial U}{\partial V}\right)_T \left(\frac{\partial V}{\partial T}\right)_P + P\left(\frac{\partial V}{\partial T}\right)_P$$

Equation [3-40] expresses the partial derivative of U with respect to V at constant T:

$$\left(\frac{\partial U}{\partial V}\right)_T = T\left(\frac{\partial P}{\partial T}\right)_V - P.$$

Substitution of this relationship into the expression for $C_P - C_V$ leads to

$$C_P - C_V = T\left(\frac{\partial P}{\partial T}\right)_V \left(\frac{\partial V}{\partial T}\right)_P.$$

Equation [2-32] defines the equation of state: $PV = nRT + nbP$.

$$\left(\frac{\partial}{\partial T}\left[\frac{nRT}{V - nb}\right]\right)_V = \frac{nR}{V - nb} \quad \text{and} \quad \left(\frac{\partial}{\partial T}\left[\frac{nRT}{P} + nb\right]\right)_P = \frac{nR}{P}.$$

$$C_P - C_V = \left(\frac{nR}{V - nb}\right)\left(\frac{nRT}{P}\right).$$

To check this result, we see if it reduces to nR in conditions where the gas should behave ideally, such as for very large volumes. If $V >> nb$, it reduces to $nR(nRT/PV) \rightarrow nR$.

21. Nitrogen would be a better choice for household uses, since its Joule-Thomson coefficient is greater than zero at 1 atm; therefore, upon expanding nitrogen gas cools. At 1 atm, the Joule-Thomson coefficient for hydrogen is less than zero; it will actually warm upon expansion.

Chapter 4

Phases and Multicomponent Systems

Exercises

1. Substitution of [4-6] and [4-7] into [4-4] gives

$$-\frac{nR\left(\frac{8a}{27bR}\right)}{(3nb-nb)^2}+\frac{2an^2}{(3nb)^3}=-\frac{8anR}{27bR(4n^2b^2)}+\frac{2an^2}{27n^3b^3}$$

$$=-\frac{2a}{27n^3b^3}+\frac{2a}{27n^3b^3}=0.$$

Making the substitutions into [4-5], we get

$$\frac{2nR\left(\frac{8a}{27bR}\right)}{(3nb-nb)^3}-\frac{6an^2}{(3nb)^4}=\frac{16anR}{27bR(8n^3b^3)}-\frac{6an^2}{81\,n^4b^4}$$

$$= \frac{2a}{27n^2b^4} - \frac{2a}{27\,n^2b^4} = 0.$$

$$P_c = \frac{nR\left(\frac{8a}{27bR}\right)}{(3nb - nb)} - \frac{an^2}{(3nb)^2} = \frac{8anR}{54bRnb} - \frac{an^2}{9n^2b^2}$$

$$= \frac{4a}{27b^2} - \frac{a}{9b^2} = \frac{a}{27b^2}.$$

2.

	a (L^2 bar mol^{-2})	b (L mol^{-1})	V_c (L)	T_c (K)
Hydrogen	0.2453	0.02651	0.07953	32.97
Water	5.537	0.03049	0.09147	647.2
Methane	2.300	0.04301	0.12903	190.6
Benzene	18.82	0.1193	0.3579	562.2

3. From [4-6], $b = V_c / 3n$. Using this in [4-7] gives us

$$a = T_c(27bR) / 8 = 9T_cV_cbR/8n.$$

Now, we substitute these expressions for a and b into [2-31]:

$$\left(P + \frac{9T_cV_cbR}{8n}\frac{n^2}{V^2}\right)\left(V - n\left(\frac{V_c}{3n}\right)\right) = nRT$$

$$\left(P + \frac{9T_cV_cnbR}{8V^2}\right)\left(\frac{3V - V_c}{3}\right) = nRT$$

$$P = \frac{3nRT}{3V - V_c} - \frac{9T_cV_cnbR}{8V^2}.$$

Although this expression represents the pressure in terms of just T, T_c, V, and V_c, another set of variables greatly simplifies this expression. *Reduced variables* are the pressure, temperature, and volume scaled by the critical values: $P_r = P / P_c$, $T_r = T / T_c$, and $V_r = V / V_c$. We can now write the van der Waals equation in terms of these variables:

$$\frac{PV_c}{T_c} = \frac{3nR(T / T_c)}{3(V / V_c) - (V_c / V_c)} - \frac{9nbR}{8(V / V_c)^2}.$$

Since $V_c / T_c = 3nb \times (27bR / 8a) = 3nR / 8\ P_c$,

$$\frac{P_r\,3nR}{8} = \frac{3nRT_r}{3V_r - 1} - \frac{9nbR}{8V_r^{\,2}}$$

$$P_r = \frac{8T_r}{3V_r - 1} - \frac{3R}{V_r^{\,2}}.$$

Note that the van der Waals equation in reduced variables has no empirical parameters; it is identical for all gases.

4. Molar volume (V_m) = volume (L) / n (mol).

$$\text{Ice: } \left(\frac{cm^3}{0.915g}\right)\left(\frac{1\ L}{1000\ cm^3}\right)\left(\frac{18.015g}{mol}\right) = 0.019688\ L\ mol^{-1}.$$

$$\text{Water: } \left(\frac{cm^3}{1.000\ g}\right)\left(\frac{1\ L}{1000\ cm^3}\right)\left(\frac{18.015g}{mol}\right) = 0.018015\ L\ mol^{-1}.$$

$$\Delta V_m = (0.018015 - 0.019688)\ (L\ mol^{-1}) = -1.673 \times 10^{-3}\ (L\ mol^{-1}).$$

5. a. One. The ammonium chloride (NH_4Cl) is in equilibrium with the ammonia (NH_3) and the hydrogen chloride (HCl), and so only the total amount of NH_4Cl can be varied.

   b. Two. Since the chlorine gas is in excess, and it is the only component in the vapor phase, the amount of sodium and the amount of chlorine must each be specified; however, the amount of sodium chloride formed will be determined by the amount of sodium and chloride.

c. Three. The vinegar will be partitioned between the oil and water phases, yet all three can be independently varied with respect to one another.

6. a. There is one component ($C = 1$) and three phases ($P = 3$). We apply the Gibbs' phase rule [4-11]: $f = 2 - 3 + 1 = 0$. There are no degrees of freedom; this system is completely specified (as we know, the triple point exists at a fixed value of T, P, V).

   b. If we describe the beverage as consisting of a syrup, water, and carbon dioxide, there are three components ($C = 3$) and two phases, liquid and vapor ($P = 2$); $f = 2 - 2 + 3 = 3$. So once the volume and pressure are fixed, only one degree of freedom is left for the bottler (either the temperature or one composition variable).

   c. $C = 1$ and $P = 1$: $f = 2 - 1 + 1 = 2$.

7. For a one-component system , [4-14] becomes

$$0 = V\,dP - S\,dT - n\,dg.$$

In Chapter 3 it was shown that at constant temperature and pressure, dG is the criterion that determines whether a process was spontaneous or not. If the same conditions are imposed, the Gibbs–Duhem equation for one component becomes simply

$$0 = n\,dg.$$

Thus, the chemical potential, g, must be constant at equilibrium.

8. Equation [4-29] relates the changes in the molar volume and the molar enthalpy for a phase transition and the temperature of the transition to the slope of the phase equilibrium line.

$$\frac{dT}{dP} = \frac{T\,\Delta V_{tr}}{\Delta H_{tr}} = \frac{(273\text{ K})(-1.67 \times 10^{-3}\text{L mol}^{-1})}{(6010\text{ J mol}^{-1})(9.869 \times 10^{-3}\text{ L atm J}^{-1})}$$

$$= -7.687 \times 10^{-3}\text{ K atm}^{-1}.$$

$$\Delta T \approx \frac{dT}{dP}\,\Delta P = (-7.687 \times 10^{-3}\ \text{K atm}^{-1})\ (0.1\ \text{bar}) \frac{0.9869\ \text{atm}}{\text{bar}}$$

$$= -7.586 \times 10^{-4}\ \text{K}.$$

9. Realizing that a liquid boils when its vapor pressure equals atmospheric pressure, we can restate the problem to ask what is the temperature at which the vapor pressure equals 0.8 bar? We can find this temperature by applying the Clausius–Clapeyron [4-34]:

$$\ln \frac{0.8\ \text{bar}}{1.0\ \text{bar}} = -\frac{10{,}000\ \text{J mol}^{-1}}{8.314\ \text{J mol}^{-1}\text{K}^{-1}} \left(\frac{1}{T} - \frac{1}{300\ \text{K}}\right).$$

$$1.8552 \times 10^{-4} = \frac{1}{T} - 3.33333 \times 10^{-3}.$$

$$T = 284.18\ \text{K}. \qquad \Delta T = 284.18 - 300.0 = -15.82\ \text{K}.$$

10. Higher. Atmospheric pressure is lower at high altitude, and this decrease in pressure will lead to the cake batter expanding faster. To prevent the cake from overexpanding, you must bake at a higher temperature which will “set” the cake faster.

Additional Exercises

11. We consider the Berthelot gas equation [2-30]. To evaluate dP/dV, we first write P in terms of all other variables:

$$P = \frac{nRT}{V - a\left(1 - \frac{b}{T^2}\right)}.$$

$$\frac{\partial P}{\partial V} = -\frac{nRT}{\left(V - a\left(1 - \frac{b}{T^2}\right)\right)^2}.$$

The only way this derivative can become equal to zero is if $T \rightarrow 0$ or $V \rightarrow \infty$. Therefore, the condition for a critical point cannot be met by this equation of state.

[2-32]:

$$P = \frac{nRT}{V - nb}.$$

$$\frac{\partial P}{\partial V} = -\frac{nRT}{(V - nb)^2}.$$

As with the Berthelot gas equation, this derivative can become zero only if $T \rightarrow 0$ or $V \rightarrow \infty$; thus, it also does not predict a critical point.

[2-29]:

$$PV = nRT + aP + bP^2 + cP^3.$$

For this equation, it is simpler first to evaluate $\partial V/\partial P$ and then to invert that expression to get $\partial P/\partial V$:

$$V = \frac{nRT}{P} + a + bP + cP^2.$$

$$\frac{\partial V}{\partial P} = -\frac{nRT}{P^2} + b + 2cP.$$

$$\frac{\partial P}{\partial V} = \frac{P^2}{-nRT + bP^2 + 2cP^3}.$$

This derivative can go to zero only if $P \rightarrow 0$ while $T \neq 0$, or if $P \rightarrow \infty$. Again, this equation cannot predict critical behavior.

[2-33]:

$$P = nRT\left(\frac{1}{V} + \frac{b}{V^2} + \frac{c}{V^3} + \frac{d}{V^4}\right).$$

$$\frac{\partial P}{\partial V} = -\frac{nRT}{V^5}\left(V^3 + 2bV^2 + 3cV + 4d\right).$$

The term in parentheses is a cubic polynomial in V, so there must be at least one value of V for which it would go to zero, and there could be three (the number of zeros will depend on the actual values of b, c, and d).

$$\frac{\partial^2 P}{\partial V^2} = \frac{2nRT}{V^6}\left(V^3 + 3bV^2 + 6cV + 10d\right).$$

Again, like the first derivative, the second derivative depends on a cubic polynomial in V. It is possible for this equation of state to predict a critical point, depending on the particular values of b, c, and d.

[2-34]:

$$P = nRT\left(\frac{1}{V} + \frac{b}{V^2} + \frac{c}{V^3} - \frac{d}{V^2T} - \frac{e}{V^3T} - \frac{f}{VT^3}\right).$$

$$\frac{\partial P}{\partial V} = -\frac{nR}{V^4T}\left(V^2T + 2bVT + 3cT - 2dV - 3e - \frac{2fV^3}{T^2}\right).$$

$$\frac{\partial^2 P}{\partial V^2} = \frac{2nR}{V^5T}\left(V^2T + 3bVT + 4cT - 3dV - 4e - \frac{3fV^3}{T^2}\right).$$

As with the previous equation, the first and second derivatives depend on a cubic polynomial in V, so it is possible for there to be a value of V and T (depending on the parameters b, c, d, e, and f) where the derivatives go to zero, and therefore this equation of state can also predict critical behavior.

12. We use [4-6] and [4-7] to evaluate V_c and T_c, and see Problem 1 for the expression for P_c. The values of V_c, T_c, and P_c are given for all the diatomic

and triatomic molecules for which van der Waals parameters are given in Table 2.2 in the following spreadsheet. All values are calculated assuming 1 mol of gas (n = 1).

	$a(L^2\ bar\ mol^{-1})$	$b\ (L\ mol^{-1})$	V_c (L)	T_c (K)	P_c (bar)
H_2	0.2453	0.02651	0.0795	32.98	12.927
N_2	1.37	0.0387	0.1161	126.16	33.879
O_2	1.382	0.03186	0.0956	154.59	50.426
Cl_2	6.343	0.05422	0.1627	416.92	79.912
H_2O	5.537	0.03049	0.0915	647.19	220.595
H_2S	4.544	0.04339	0.1302	373.22	89.391
CO	1.472	0.03948	0.1184	132.88	34.978
CO_2	3.658	0.04268	0.1280	305.45	74.376
CS_2	11.25	0.07262	0.2179	552.09	79.009
NO	1.46	0.0289	0.0867	180.04	64.743
NO_2	5.36	0.0443	0.1329	431.20	101.156
N_2O	3.852	0.04435	0.1331	309.53	72.533

13. For ammonia (NH_3), a = 4.225 L^2 bar mol^{-2} and b = 0.03713 L mol^{-1}.

$$T_c = \frac{8 \times 4.225\ L^2\ bar\ mol^{-2}}{27 \times 0.03713\ L\ mol^{-1} \times 8.314 \times 10^{-2}\ L\ bar\ mol^{-1}\ K^{-1}}$$

$$= 405.5\ K.$$

Now, we use this temperature and these parameters in the van der Waals equations to get the pressure.

$$P = \frac{(1)(8.314 \times 10^{-2}\ \frac{L\ bar}{mol\ K})(405.5\ K)}{(V\ [L] - (1\ mol)(0.03713\ L\ mol^{-1})} - \frac{4.225 \frac{L^2\ bar}{mol^2}(1\ mol)^2}{(V\ [L])^2}$$

The value of the pressure for V = 0.01 L to 0.20 L is given below, and plotted as well. The inflection point, where the slope of the line equals zero, (indicative of the critical point) is clearly seen in the plot.

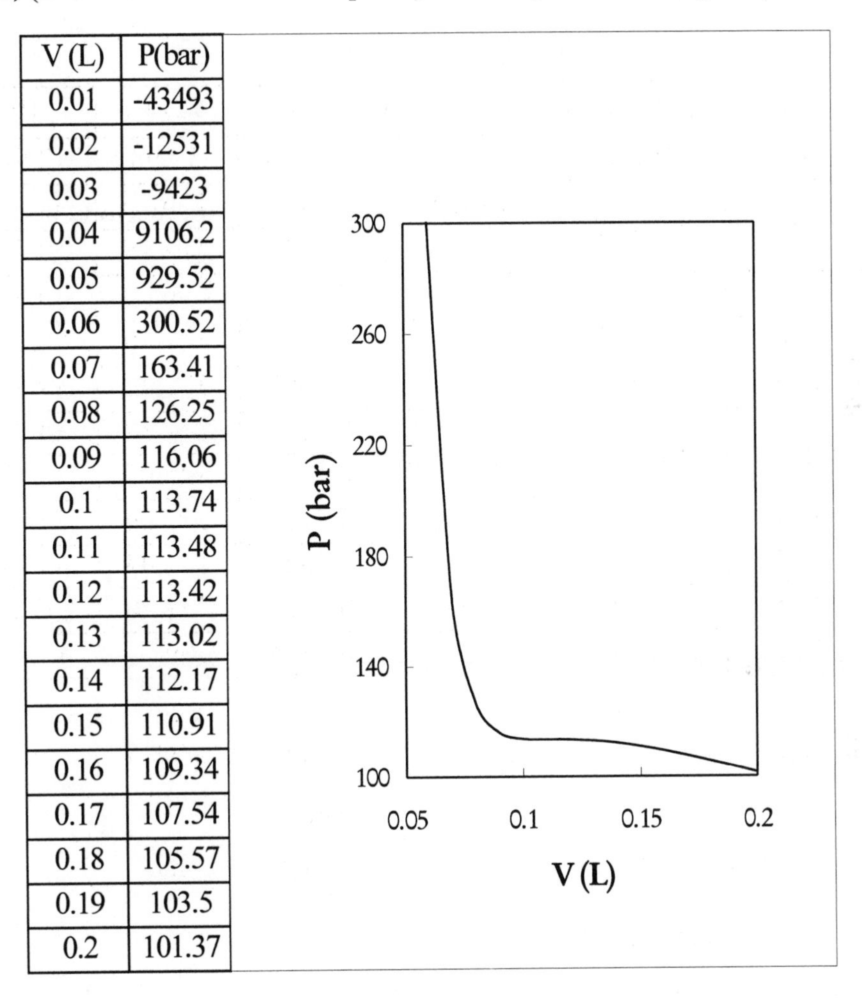

V (L)	P(bar)
0.01	-43493
0.02	-12531
0.03	-9423
0.04	9106.2
0.05	929.52
0.06	300.52
0.07	163.41
0.08	126.25
0.09	116.06
0.1	113.74
0.11	113.48
0.12	113.42
0.13	113.02
0.14	112.17
0.15	110.91
0.16	109.34
0.17	107.54
0.18	105.57
0.19	103.5
0.2	101.37

14. If we start with [4-12],

$$dG = V\,dP - S\,dT + \sum_i g_i\,dn_i\,.$$

Now, if we divide by dn_i and impose constant pressure, temperature, and $n_{j\neq i}$, we arrive at

$$\left(\frac{\partial G}{\partial n_i}\right)_{T,P,n_j} = g_i .$$

Since G = H – TS, then H = G + TS, and so

$$dH = V\,dP + T\,dS + \sum_i g_i\, dn_i .$$

If we now divide by dn_i and impose constant pressure, entropy and $n_{j \neq i}$, the partial molar enthalpy can be seen to be

$$\left(\frac{\partial H}{\partial n_i}\right)_{S,P,n_j} = g_i .$$

Similarly, since G = A + PV, then A = G – PV, and so

$$dA = -P\,dV - S\,dT + \sum_i g_i\, dn_i$$

If we now divide by dn_i and impose constant volume, temperature and $n_{j \neq i}$, the partial molar Helmholtz free energy is

$$\left(\frac{\partial A}{\partial n_i}\right)_{T,V,n_j} = g_i .$$

Finally, U = H – PV; therefore,

$$dU = -P\,dV + T\,dS + \sum_i g_i\, dn_i .$$

If we now divide by dn_i and impose constant volume, entropy and $n_{j \neq i}$, the partial molar energy can be seen to be

$$\left(\frac{\partial U}{\partial n_i}\right)_{S,V,n_j} = g_i .$$

Since all these partial molar quantities equal g_i, they must be equal to one another.

15. $\ln\left(\frac{T_2}{T_1}\right) = \ln\left(\frac{T_1 + T_2 - T_1}{T_1}\right) = \ln\left(1 + \frac{T_2 - T_1}{T_1}\right)$.

The series expansion for ln (1 + x), when truncated at first order [see Appendix I] is

$$\ln(1 + x) \approx x,$$

and therefore

$$\ln\left(\frac{T_2}{T_1}\right) = \ln\left(1 + \frac{T_2 - T_1}{T_1}\right) \approx \frac{T_2 - T_1}{T_1}.$$

Turning to Example 4.1, we can now write

$$P_2 - P_1 \approx \frac{\Delta\overline{H}_{tr}}{\Delta\overline{V}_{tr}} \frac{T_2 - T_1}{T_1}.$$

We can rearrange this expression and write it in terms of changes in P and T:

$$\frac{\Delta P}{\Delta T} \approx \frac{\Delta\overline{H}_{tr}}{T_{tr}\,\Delta\overline{V}_{tr}}.$$

In general, this expression does not indicate that the phase boundary curve between solids and liquids is a straight line, because the slope ($\Delta P/\Delta T$) depends on T. However, since the magnitude of ΔH is typically much greater than that of T ΔV, the slope is nearly vertical. Thus, the melting temperature changes very little, and when ΔH and ΔV remain nearly constant, it is appropriate to approximate the solid–liquid boundary as a straight line (see Figures 4.1 and 4.3, for example).

16. According to [4-34] and [4-35], if the enthalpy change were constant with temperature, a plot of ln (P) vs. 1/T would be a linear plot. Alternatively, the change in enthalpy would be the same for any pair of data points.

If the change in enthalpy were not constant with temperature, then the plot of ln (P) vs. 1/T would not be linear, and the value of the enthalpy change would change as different data points were used to evaluate it.

17. $$dP = \frac{\Delta\overline{H}_{tr}}{\Delta\overline{V}_{tr}}\frac{dT}{T} = \frac{1}{\Delta\overline{V}_{tr}}\left(\frac{A + BT + CT^2}{T}\right)dt = \frac{1}{\Delta\overline{V}_{tr}}\left(\frac{A}{T} + B + CT\right)dt.$$

$$\int_{P_1}^{P_2} dP = \int_{T_1}^{T_2} \frac{1}{\Delta\overline{V}_{tr}}\left(\frac{A}{T} + B + CT\right)dt.$$

$$P_2 - P_1 = \frac{1}{\Delta\overline{V}_{tr}}\left(A\ln(T) + BT + \frac{1}{2}CT^2\right)\Bigg|_{T_1}^{T_2}$$

$$P_2 - P_1 = \frac{1}{\Delta\overline{V}_{tr}}\left(A\ln\left(\frac{T_2}{T_1}\right) + B(T_2 - T_1) + \frac{1}{2}C({T_2}^2 - {T_1}^2)\right).$$

18. The most significant difference between the rare gas atoms (Ne and Ar) and the two halogen molecules (F_2 and Cl_2) is the lower symmetry of the halogen molecules. Because the atoms are spherically symmetric, they have no nonzero electrical moments; the only attractive interactions arise from weak dispersion forces (instantaneous dipole–dipole interactions). The halogen molecules are linear; therefore, they have sizable electrical quadrupole moments, which can have relatively much greater attractions than arise from dispersion forces. And since Cl_2 is significantly larger than F_2, its quadrupole moment should be likewise larger, causing it to exert stronger attractive forces than F_2. Alternatively, the magnitude of the dispersion forces that Ne and Ar can exert on other Ne and Ar atoms is much less dependent on atomic size.

19. A plot of T_b (K) vs. ΔH_{vap} (kJ) for the nonmetallic species listed in Table 4.1 is given below.

For a phase change, $\Delta G = 0$ (both phases are in equilibrium at the transition temperature) and so $\Delta H_{tr} - T_{tr}\,\Delta S_{tr} = 0$. Therefore, $\Delta S_{tr} = \Delta H_{tr} / T_{tr}$. This plot indicates that, for these nonmetallic compounds, the value of ΔS_{tr} is roughly constant (since the plot is linear in T_b vs. ΔH_{vap}).

If we consider the vaporization process, the greatest contribution to the total change in entropy comes from the entropy of the gas phase; the difference in entropy between one liquid and another is so much smaller that the overall ΔS is roughly the same for all these compounds.

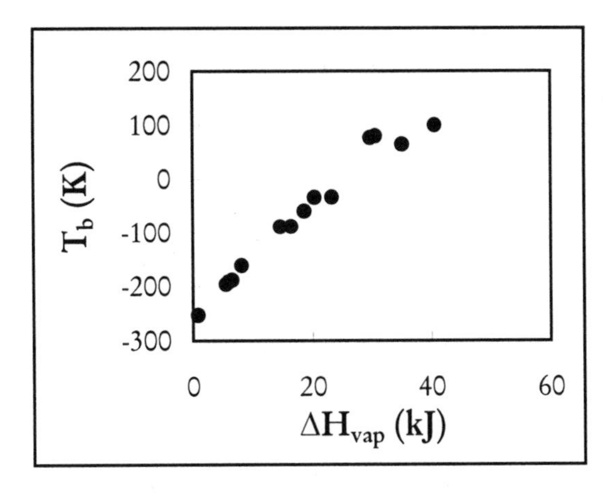

This phenomenon was observed and described by Trouton; Trouton's rule says that ΔH_{vap} (J) / T_b (K) $\approx$ 88 J K^{-1} mol^{-1}.

20.

H_2O		H_2S	
T (°C)	P_{vap} (bar)	T (°C)	P_{vap} (bar)
100.000	1.000	-59.550	1.000
110.000	1.408	-49.550	1.601
120.000	1.949	-39.550	2.463
130.000	2.653	-29.550	3.656
140.000	3.559	-19.550	5.261
150.000	4.709	-9.550	7.365
160.000	6.150	0.450	10.058
170.000	7.935	10.450	13.438
180.000	10.124	20.450	17.604
190.000	12.782	30.450	22.653
200.000	15.979	40.450	28.685

Even though water has the higher ΔH_{vap} (40.65 kJ vs. 18.67 kJ), the inverse dependence of the vapor pressure on the boiling point and the much lower boiling point of H_2S leads to a greater increase in the vapor pressure of H_2S than that of H_2O.

Chapter 5

Activity and Equilibrium of Gases and Solutions

Exercises

1. $V_{ideal} = (1\ \text{mol})(8.314 \times 10^{-2}\ \text{L bar/mol K})(500\ \text{K}) / (5\ \text{bar}) = 8.314\ \text{L}$.

   For neon, $a = 0.208\ (\text{L}^2\ \text{bar mol}^{-1})$ and $b = 0.01672\ (\text{L mol}^{-1})$.

$$P_{Ne} = \frac{(1\ \text{mol})(8.314 \times 10^{-2}\ \frac{\text{L bar}}{\text{K mol}})(500\ \text{K})}{8.314\ \text{L} - (1\ \text{mol})(0.01672\ \text{L mol}^{-1})} - \frac{(0.208\frac{\text{L}^2\ \text{bar}}{\text{mol}^2})(1\ \text{mol})^2}{(8.314\ \text{L})^2}$$

$$= 5.007\ \text{bar}.$$

   For mercury, $a = 5.193\ (\text{L}^2\ \text{bar mol}^{-1})$ and $b = 0.01057\ (\text{L mol}^{-1})$.

$$P_{Hg} = \frac{(1\ \text{mol})(8.314 \times 10^{-2}\ \frac{\text{L bar}}{\text{K mol}})(500\ \text{K})}{8.314\ \text{L} - (1\ \text{mol})(0.01057\ \text{L mol}^{-1})} - \frac{(5.193\frac{\text{L}^2\ \text{bar}}{\text{mol}^2})(1\ \text{mol})^2}{(8.314\ \text{L})^2}$$

$$= 4.931\ \text{bar}.$$

For benzene, a = 18.82 (L^2 bar mol^{-1}) and b = 0.1193 (L mol^{-1}).

$$P_{benzene} = \frac{(1\text{ mol})(8.314\times10^{-2}\ \frac{\text{L bar}}{\text{K mol}})(500\text{ K})}{8.314\text{ L}-(1\text{ mol})(0.1193\text{ L mol}^{-1})} - \frac{(18.82\,\frac{\text{L}^2\text{ bar}}{\text{mol}^2})(1\text{ mol})^2}{(8.314\text{ L})^2}$$

$$= 4.800\text{ bar.}$$

2. Ideal gas: $PV - nRT = 0$.

   [2-29]: $PV - nRT = c_1 P + c_2 P^2 + c_3 P^3$.

   [2-29] – ideal gas = $c_1 P + c_2 P^2 + c_3 P^3$, which obviously goes to zero as P goes to zero.

   [2-30]: $PV - nRT = aP\left(1-\frac{b}{T^2}\right)$.

   [2-30] – ideal gas = $aP\left(1-\frac{b}{T^2}\right)$, again, this value goes to zero as P goes to zero.

   [2-31]: $PV - nRT = nbP - \left(\frac{an^2}{V^2}\right)(V-nb)$.

   [2-31] – ideal gas = $nbP - \left(\frac{an^2}{V^2}\right)(V-nb)$.

   As $P \to 0$, $V \to \infty$ (when temperature is held constant). Thus, the first term goes directly to zero as $P \to 0$, and the second term goes to zero as $V \to \infty$.

3. If we assume that the vapor phase behaves ideally (which is reasonable at these low pressures), the activity is the ratio of the partial pressure of the mixture to the vapor pressure of the pure liquid [5-30]:

$$a_{alcohol} = \frac{0.05\text{ bar}}{0.08\text{ bar}} = 0.625; \qquad a_{water} = \frac{0.02\text{ bar}}{0.03\text{ bar}} = 0.667.$$

The activity coefficients are the ratio of the activity to the mole fraction [5-35]:

$$\gamma_{\text{alcohol}} = \frac{0.625}{0.5} = 1.25; \qquad \gamma_{\text{water}} = \frac{0.667}{0.5} = 1.333.$$

4. If we assume ideal behavior of the vapor phase, the mole fraction of component 1 in the vapor phase is given by

$$y_1 = P_1 / P^{tot}$$

If we also assume ideal solution behavior (i.e., Raoult's law applies),

$$P_1 = x_1 P_1^{pure}.$$

We can now apply these two relationships to find y_1 in terms of x_1:

$$P^{tot} = P_1 + P_2 = x_1 P_1^{pure} + x_2 P_2^{pure} = x_1 P_1^{pure} + (1 - x_1) 3 P_1^{pure}$$

$$= 3 P_1^{pure} - 2 x_1 P_1^{pure} = (3 - 2 x_1) P_1^{pure},$$

where we used the information that $3 P_1^{pure} = P_2^{pure}$. Finally,

$$y_1 = x_1 P_1^{pure} / (3 - 2 x_1) P_1^{pure} = x_1 / (3 - 2 x_1).$$

$$y_1(x_1 = 0.25) = 0.25 / (3 - (2 \times 0.25)) = 0.10.$$

$$y_1(x_1 = 0.50) = 0.25 / (3 - (2 \times 0.50)) = 0.25.$$

$$y_1(x_1 = 0.75) = 0.25 / (3 - (2 \times 0.75)) = 0.50.$$

5. For the melting points (remember to use $-\Delta H_{fusion}$ in [5-58]):

$$H_2O:\ T_2 = \left(\frac{-(8.314\ \text{J mol}^{-1}\text{K}^{-1})(0.01)}{-6010\ \text{J mol}^{-1}} + \frac{1}{273.15\ \text{K}} \right)^{-1} = 272.121\ \text{K}.$$

$$\Delta T_m = 272.121 - 273.15 = -1.027\ \text{K}.$$

$$H_2S:\ T_2 = \left(\frac{-(8.314\ \text{J mol}^{-1}\text{K}^{-1})(0.01)}{-23{,}800\ \text{J mol}^{-1}} + \frac{1}{187.65\ \text{K}} \right)^{-1} = 187.527\ \text{K}.$$

$$\Delta T_m = 187.527 - 18765 = -0.123\ \text{K}.$$

$$CCl_4:\ T_2 = \left(\frac{-(8.314\ \text{J mol}^{-1}\text{K}^{-1})(0.01)}{-3280\ \text{J mol}^{-1}} + \frac{1}{250.15\ \text{K}} \right)^{-1} = 248.576\ \text{K}.$$

$$\Delta T_m = 248.576 - 250.15 = -1.574\ \text{K}.$$

$$CH_3OH:\ T_2 = \left(\frac{-(8.314\ \text{J mol}^{-1}\text{K}^{-1})(0.01)}{-3180\ \text{J mol}^{-1}} + \frac{1}{175.47\ \text{K}} \right)^{-1} = 174.669\ \text{K}.$$

$$\Delta T_m = 174.669 - 175.47 = -0.801\ \text{K}.$$

$$C_6H_6:\ T_2 = \left(\frac{-(8.314\ \text{J mol}^{-1}\text{K}^{-1})(0.01)}{-9950\ \text{J mol}^{-1}} + \frac{1}{278.68\ \text{K}} \right)^{-1} = 278.033\ \text{K}.$$

$$\Delta T_m = 278.033 - 278.68 = -0.647\ \text{K}.$$

For the boiling points:

$$H_2O:\ T_2 = \left(\frac{-(8.314\ \text{J mol}^{-1}\text{K}^{-1})(0.01)}{40{,}650\ \text{J mol}^{-1}} + \frac{1}{373.15\ \text{K}} \right)^{-1} = 373.435\ \text{K}.$$

$$\Delta T_b = 373.435 - 373.15 = 0.285\ \text{K}.$$

$$H_2S:\ T_2 = \left(\frac{-(8.314\ \text{J mol}^{-1}\text{K}^{-1})(0.01)}{18{,}670\ \text{J mol}^{-1}} + \frac{1}{213.6\ \text{K}} \right)^{-1} = 213.803\ \text{K}.$$

$$\Delta T_b = 213.803 - 213.6 = 0.203 \text{ K.}$$

$$CCl_4: \; T_2 = \left(\frac{-(8.314 \text{ J mol}^{-1}\text{K}^{-1})(0.01)}{29{,}820 \text{ J mol}^{-1}} + \frac{1}{349.95 \text{ K}} \right)^{-1} = 350.291 \text{ K}.$$

$$\Delta T_b = 350.291 - 349.95 = 0.341 \text{ K.}$$

$$CH_3OH: \; T_2 = \left(\frac{-(8.314 \text{ J mol}^{-1}\text{K}^{-1})(0.01)}{35{,}210 \text{ J mol}^{-1}} + \frac{1}{337.75 \text{ K}} \right)^{-1} = 338.020 \text{ K}.$$

$$\Delta T_b = 338.020 - 337.75 = 0.270 \text{ K.}$$

$$C_6H_6: \; T_2 = \left(\frac{-(8.314 \text{ J mol}^{-1}\text{K}^{-1})(0.01)}{30{,}720 \text{ J mol}^{-1}} + \frac{1}{353.24 \text{ K}} \right)^{-1} = 353.578 \text{ K}.$$

$$\Delta T_b = 353.578 - 353.24 = 0.338 \text{ K.}$$

6. From Table 4.1, T_m for gold is 1064.18 + 273.15 = 1337.33 K, and ΔH_{fusion} = 12.55 kJ mol^{-1}. Using [5-58], we obtain

$$-X_{Ag} = \frac{-12{,}550 \text{ J mol}^{-1}}{8.314 \text{ J mol}^{-1} \text{ K}^{-1}} \left(\frac{1}{1237.33} - \frac{1}{1337.33} \right) = -0.091.$$

$$X_{Ag} = 0.091.$$

Additional Exercises

---

7. As described in Chapter 2 [see pp. 47-48], the equation of state of a gas can be described in terms of the quantity $X = PV - nRT$. X can be expanded in a general power series where P, V, and T are the variables (see p. 48). Let's choose to truncate the power series at the first-order terms:

$$X = PV - nRT = c_{000} + c_{100}P + c_{010}V + c_{001}T$$

We would determine the coefficients, c_{000}, c_{100}, c_{010}, and c_{001} by fitting experimentally determined values of X(P,V,T) to this expression. As P → 0, this expression reduces to

$$X = c_{000} + c_{010}V + c_{001}T.$$

At a fixed temperature, as the pressure decreases we know that the volume will increase, so X will not go to zero, and this equation of state does not reduce to ideal behavior, as a proper equation of state must.

The problem is that we know that the equation of state must depend on pressure and volume inversely; a physically realistic expansion would use P and (1/V) as expansion variables.

8. We use [5-18] to find the fugacity:

$$RT\ln\frac{f}{P} = \int_0^P (\overline{V}_{real} - \overline{V}_{ideal})\,dP = \int_0^P \left[\left(\frac{RT}{P} + b\right) - \frac{RT}{P}\right] dP = bP$$

$$f = P\,e^{(bP/RT)} = 1(\text{bar})\,e^{-\left(\frac{0.017\ \text{L mol}^{-1} \times 1\ \text{bar}}{0.08314\ \text{L bar mol}^{-1}\,\text{K}^{-1} \times 400\ \text{K}}\right)}$$

$$= 1.0005\ \text{bar}.$$

9. We start with [5-6]:

$$\ln a = \frac{1}{RT}\int_1^P \overline{V}\,dp = \frac{1}{RT}\int_1^P \left[\frac{RT}{P} + c\left(1 - \frac{d}{T^2}\right)\right] dP = [\ln P]_1^P + \frac{c}{RT}\left(1 - \frac{d}{T^2}\right)[P]_1^P$$

$$= \ln P + \frac{c}{RT}\left(1 - \frac{d}{T^2}\right)(P - 1).$$

10. $$Z = \frac{PV}{nRT} = 1 + \frac{c_1 P}{nRT} + \frac{c_2 P^2}{nRT} + \frac{c_3 P^3}{nRT}.$$

11. $\frac{V}{n} = \overline{V} = \frac{ZRT}{P}$; $\overline{V}_{ideal} = \frac{Z_{ideal}RT}{P} = \frac{RT}{P}$.

Therefore, $\overline{V} - \overline{V}_{ideal} = \frac{ZRT}{P} - \frac{RT}{P} = \frac{RT}{P}(Z-1)$.

We can now write [5-18] as

$$RT \ln \frac{f}{P} = RT\int_0^P \left(\frac{Z-1}{P}\right) dP.$$

12. For [2-32], PV = n(RT + bP), the compressibility factor is

$$Z = \frac{PV}{nRT} = 1 + \frac{bP}{RT}.$$

Now, we can apply the relationship from the last problem to find the fugacity.

$$RT \ln \frac{f}{P} = RT\int_0^P \left(\frac{Z-1}{P}\right) dP = RT\int_0^P \left[\frac{1}{P}\left(1 + \frac{bP}{RT} - 1\right)\right] dP = RT \times \frac{bP}{RT}$$

$$\ln \frac{f}{P} = \frac{bP}{RT}$$

$$f = P\,e^{(bP/RT)} = 100\ (\text{bar})\ e^{(0.050\ \text{L mol}^{-1} \times 100\ \text{bar}\ /\ (0.08314 \times 400\ \text{K})}$$

$$= 116\ \text{bar}.$$

At P = 1 bar,

$$f = P\,e^{(bP/RT)} = 1\ (\text{bar})\ e^{(0.050\ \text{L mol}^{-1} \times 1\ \text{bar}\ /\ (0.08314 \times 400\ \text{K})}$$

$$= 1.0015\ \text{bar}.$$

13. For a van der Waal's gas, expressed in terms of molar volume,

$$P = \frac{RT}{\overline{V} - b} - \frac{a}{\overline{V}^2}.$$

Because it is difficult to write V(P,T) to substitute into [5-18], instead we write dP in terms of dV. If T is held constant, then

$$dP = \left(-\frac{RT}{(\overline{V}-b)^2} + \frac{2a}{\overline{V}^3} \right) d\overline{V}.$$

For an ideal gas, since $P = RT/\overline{V}$,

$$dP = -\frac{RT}{\overline{V}^2}\, d\overline{V}.$$

Now, we can use [5-18] to find the fugacity coefficient, $\gamma = f/P$.

$$RT \ln \gamma = \int_0^P \left(\overline{V}_{real} - \overline{V}_{ideal}\right) dP = \int_0^P \overline{V}_{real}\, dP - \int_0^P \overline{V}_{ideal}\, dP$$

$$= \int_\infty^{\overline{V}} \overline{V}_{vdw} \left(-\frac{RT}{(\overline{V}-b)^2} + \frac{2a}{\overline{V}^3} \right) d\overline{V} - \int_\infty^{\overline{V}} \overline{V}_{ideal} \left(-\frac{RT}{\overline{V}^2} \right) d\overline{V}$$

[The limits of the integration must be expressed in terms of volumes, and note that as $P \to 0$, $V \to \infty$.]

$$= -\int_\infty^{\overline{V}_{vdw}} \left(\frac{\overline{V}RT}{(\overline{V}-b)^2} \right) d\overline{V} + \int_\infty^{\overline{V}_{vdw}} \left(\frac{2a}{\overline{V}^2} \right) d\overline{V} + \int_\infty^{\overline{V}_{ideal}} \frac{RT}{\overline{V}}\, d\overline{V}$$

$$= -\int_\infty^{\overline{V}_{vdw}+b} \left(\frac{(\overline{V}'+b)RT}{\overline{V}'^2} \right) d\overline{V}' + \int_\infty^{\overline{V}_{vdw}} \left(\frac{2a}{\overline{V}^2} \right) d\overline{V} + \int_\infty^{\overline{V}_{ideal}} \frac{RT}{\overline{V}}\, d\overline{V}$$

$$= -\int_\infty^{\overline{V}_{vdw}+b} \left(\frac{RT}{\overline{V}'} \right) d\overline{V}' - \int_\infty^{\overline{V}_{vdw}+b} \left(\frac{bRT}{\overline{V}'^2} \right) d\overline{V}' + \int_\infty^{\overline{V}_{vdw}} \left(\frac{2a}{\overline{V}^2} \right) d\overline{V} + \int_\infty^{\overline{V}_{ideal}} \frac{RT}{\overline{V}}\, d\overline{V}$$

$$= -RT \ln \overline{V}'\Big|_\infty^{\overline{V}'_{vdw}+b} + \frac{bRT}{\overline{V}'}\Big|_\infty^{\overline{V}_{vdw}+b} - \frac{2a}{\overline{V}}\Big|_\infty^{\overline{V}_{vdw}} + RT \ln \overline{V}_{ideal}\Big|_\infty^{\overline{V}_{ideal}}$$

$$= RT \ln \frac{\overline{V}_{ideal}}{\overline{V}_{vdw}+b} + \frac{bRT}{\overline{V}_{vdw}+b} - \frac{2a}{\overline{V}_{vdw}}$$

$$= RT \ln \frac{RT}{P(\overline{V}_{vdw} + b)} + \frac{bRT}{\overline{V}_{vdw} + b} - \frac{2a}{\overline{V}_{vdw}}.$$

In the evaluation of the integrals, the two terms that go to infinity are of opposite sign and so cancel each other out. The final result is

$$\ln \gamma = \ln \frac{RT}{P(\overline{V}_{vdw} + b)} + \frac{b}{\overline{V}_{vdw} + b} - \frac{2a}{RT\overline{V}_{vdw}}.$$

14. The solid line indicates Raoult's law behavior ($P_1 = X_1P_1°$), and the broken line Henry's law behavior ($P_1 = X_1K_1$). Note that Raoult's law is only applicable when $X_1 \approx 1$, and Henry's law when $X_1 \approx 0$. The Henry's law constant can be found by determining the derivative of P_1 at $X_1 = 0$ (for this example, the Henry's Law constant = 31).

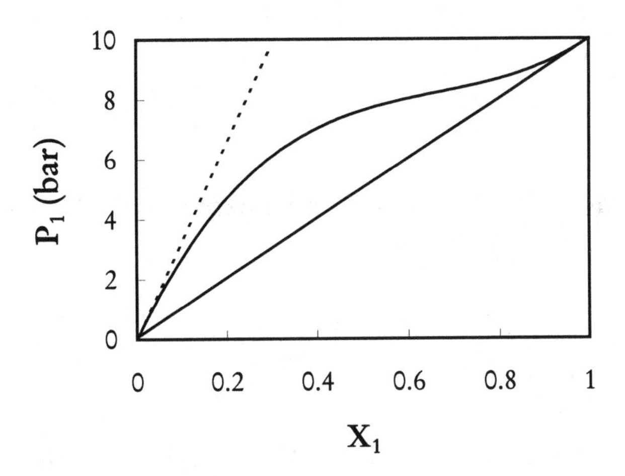

Since the activity is the ratio of the partial pressure to the vapor pressure [5-30], when $X_1 = 0.5$

$$a_1 = \frac{P_1}{P_1^{pure}} = \frac{(10 \text{ bar})(0.5)(1 + 2.1 \times (1.0 - 0.5)^2)}{(10 \text{ bar})(1.0)(1 + 2.1 \times (1.0 - 1.0)^2)} = \frac{7.625}{10} = 0.7625.$$

15. Tap water normally contains dissolved gases. The water is *saturated* when the greatest amount of gas is dissolved in water; the amount of gas in

saturated water is determined by the solubility of the gas. When the water is heated, the solubility of a gas decreases, and the excess gas forms bubbles that escape the liquid phase. The same phenomenon occurs when the water freezes, except that the bubbles cannot escape from the solid ice and are trapped.

Henry's law says that the partial pressure of the vapor of a substance in equilibrium with the substance dissolved in a liquid is proportional to the concentration of the substance in the solution. For a given quantity of water in contact with a gas at a given pressure, the mole fraction of the gas is given by the ratio of the pressure to the Henry's law constant for the gas. From the mole fraction, the number of moles of gas dissolved can be determined, and then an equation of state can be used to predict the volume of the gas when it forms bubbles.

16. Substituting $\Delta H = a + bT + cT^2$ into [5-49] and integrating, we get

$$d \ln K = \frac{\Delta H}{RT^2}\, dT = \frac{a + bT + cT^2}{RT^2}\, dT$$

$$\int_{K_1}^{K_2} d \ln K = \int_{T_1}^{T_2} \left(\frac{a}{RT^2} + \frac{b}{RT} + \frac{c}{R} \right) dT$$

$$\ln \frac{K_2}{K_1} = -\frac{a}{R}\left(\frac{1}{T_2} - \frac{1}{T_1} \right) + \frac{b}{R} \ln \frac{T_2}{T_1} + \frac{c}{R}(T_2 - T_1).$$

If we compare this result with [5-50], we see that the first term is the same; the additional terms result from the temperature dependence of ΔH.

17. We recall [2-37], $H = U + PV$. For a finite change at constant pressure,

$$\Delta U_{mix} = \Delta H_{mix} - P\, \Delta V_{mix}$$

$$= -RT^2 \sum_i^c n_i \left(\frac{\partial \ln \gamma_i}{\partial T} \right)_{P,X} - PRT \sum_i^c n_i \left(\frac{\partial \ln \gamma_i}{\partial P} \right)_{T,X}$$

$$\Delta C_{P,mix} = \left(\frac{\partial \Delta H_{mix}}{\partial T}\right)_P$$

$$= \frac{\partial}{\partial T}\left(-RT^2\sum_i^c n_i\left(\frac{\partial \ln \gamma_i}{\partial T}\right)_{P,X}\right)_P.$$

$$= -2RT\sum_i^c n_i\left(\frac{\partial \ln \gamma_i}{\partial T}\right)_{P,X} - RT^2\sum_i^c n_i\left(\frac{\partial^2 \ln \gamma_i}{\partial T^2}\right)_{P,X}.$$

For ideal solutions, the activity coefficients (γ_i) are equal to 1, and so any derivatives of γ_i will be identically zero. Thus $\Delta V_{mix} = \Delta H_{mix} = \Delta C_{P,mix} = \Delta U_{mix} = 0$ in the limit of ideal solutions. However, ΔG_{mix} and ΔS_{mix} are nonzero even for ideal solutions.

$$\Delta S_{mix}(\text{ideal}) = -\sum_i^c n_i R \ln X_i \gamma_i.$$

18.

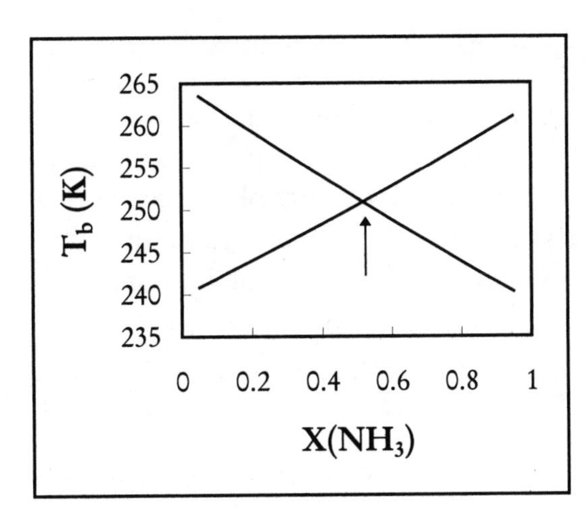

The arrow indicates the eutectic point, the solution that has the highest boiling point. It occurs at $X(NH_3) = 0.517$.

19. Assuming that the impurities are present as 1g kg^{-1}, and further that the molecular weight of the impurities is 100 g mol^{-1}, then we can find the mole fraction of the impurities:

$$X_A = \frac{1\ \text{g} / 100\ \text{g mol}^{-1}}{(1\ \text{g} / 100\ \text{g mol}^{-1} + 999\ \text{g} / 18\ \text{g mol}^{-1})} \approx 2 \times 10^{-4}.$$

For water, the molar volume is

$$\overline{V} = \left(\frac{1\ \text{ml}}{1\ \text{g}}\right)\left(\frac{1\ \text{L}}{1000\ \text{ml}}\right)\left(\frac{18\ \text{g}}{\text{mol}}\right) = 0.018\ \text{L mol}^{-1}.$$

The osmotic pressure is thus

$$\Pi = (0.082\ \text{L atm mol}^{-1}\ \text{K}^{-1})(300\ \text{K})(2 \times 10^{-4}) / (0.018\ \text{L mol}^{-1})$$

$$= 0.27\ \text{atm}.$$

To compare this value with the pressure in a typical water system, we convert atm to lb in.^{2}.

$$0.27\ \text{atm} \times 14.7\ \text{lb in.}^{-2}\ \text{atm}^{-1} \sim 4\ \text{lb in.}^{-2}.$$

Here in Grove City, the typical pressure in the water system is 70 lb in.$^{-2}$, according to city engineers. This is a gauge pressure, i.e., the pressure above atmospheric pressure. To accomplish reverse osmosis, a pressure equal to the gauge pressure of the water system, plus the osmotic pressure, would have to be applied, in this case about 75 lb in.$^{-2}$.

Chapter 6

Chemical Reactions: Kinetics, Dynamics, and Equilibrium

Exercises

1. The mean free path, λ, is the ratio of the average speed to the rate of collisions [6-8], and the rate of collisions is given by [6-3]

$$\lambda = \frac{\langle s \rangle}{z} = \frac{\langle s \rangle}{\pi R^2 D \langle s \rangle} = \frac{1}{\pi R^2 D}.$$

We assume that at the conditions stated the gases behave ideally; we can find the density from the ideal gas law.

$$D\ (\text{mol L}^{-1}) = \frac{P}{RT}$$

$D\ (He) = (1\ \text{bar}) / [(0.08314\ \text{L bar mol}^{-1}\ \text{K}^{-1})(300\ \text{K})]$

$D\,(\text{He}) = 0.04009 \text{ mol L}^{-1}$.

$$D\,(\text{He}) = \left(\frac{0.04009 \text{ mol}}{\text{L}}\right)\left(\frac{1 \text{ L}}{1 \text{ dm}^3}\right)\left(\frac{1 \text{ dm}}{10^9 \text{ Å}}\right)^3\left(\frac{6.022 \times 10^{23} \text{ molecules}}{\text{mol}}\right)$$

$$= 2.4143 \times 10^{-5} \text{ molecules Å}^{-3}.$$

$\lambda\,(\text{He}) = 1 / [\pi \times (1.2 \text{ Å})^2 \times 2.4143 \times 10^{-5} \text{ Å}^{-3}] = 9156 \text{ Å}$.

$\lambda\,(\text{Ar}) = 1 / [\pi \times (1.9 \text{ Å})^2 \times 2.4143 \times 10^{-5} \text{ Å}^{-3}] = 3653 \text{ Å}$.

2. The ratio of the constant-pressure heat capacity to the constant-volume heat capacity (C_P/C_V) for argon is 5/3. From [6-9], the speed of sound for ^{36}Ar is

$$s_{sound}(250 \text{ K}) = \sqrt{\frac{(5)(1.381\times10^{-23} \text{J K}^{-1})(250 \text{ K})}{(3)(35.967 \text{ amu}\times1.661\times10^{-27} \text{ kg amu}^{-1})}} = 310.4 \text{ m s}^{-1}.$$

$$s_{sound}(310 \text{ K}) = \sqrt{\frac{(5)(1.381\times10^{-23} \text{J K}^{-1})(310 \text{ K})}{(3)(35.967 \text{ amu}\times1.661\times10^{-27} \text{ kg amu}^{-1})}} = 345.6 \text{ m s}^{-1}.$$

For ^{40}Ar, the speed of sound is

$$s_{sound}(250 \text{ K}) = \sqrt{\frac{(5)(1.381\times10^{-23} \text{J K}^{-1})(250 \text{ K})}{(3)(39.962 \text{ amu}\times1.661\times10^{-27} \text{ kg amu}^{-1})}} = 294.4 \text{ m s}^{-1}.$$

$$s_{sound}(310 \text{ K}) = \sqrt{\frac{(5)(1.381\times10^{-23} \text{J K}^{-1})(310 \text{ K})}{(3)(39.962 \text{ amu}\times1.661\times10^{-27} \text{ kg amu}^{-1})}} = 327.9 \text{ m s}^{-1}.$$

3. Starting with [6-40] for the rate constant, k, we derive an expression for the ratio of rate constants at different temperatures:

$$\frac{k(T_2)}{k(T_1)} = \frac{Ae^{-E_{act}/RT_2}}{Ae^{-E_{act}/RT_1}} = e^{-\left(\frac{E_{act}}{R}\right)\left(\frac{1}{T_2} - \frac{1}{T_1}\right)}$$

Now, we set the ratio of the rate constants to 2 (the condition where the reaction rate doubles), to get

$$E_{act} = \frac{-R \ln 2}{\left(\frac{1}{T_2} - \frac{1}{T_1}\right)}$$

This expression relates the energy of activation to the two temperatures between which the reaction rate doubles.

a. E_{act} (T_1 = 10 K, T_2 = 20 K) = 115.3 J mol^{-1}.

b. E_{act} (T_1 = 100 K, T_2 = 110 K) = 6.340 kJ mol^{-1}.

c. E_{act} (T_1 = 300 K, T_2 = 310 K) = 53.60 kJ mol^{-1}.

d. E_{act} (T_1 = 1000 K, T_2 = 1010 K) = 582.1 kJ mol^{-1}.

4. We apply [6-43] to find $\Delta H^o_{rxn,298}$:

a. $$\Delta H^o_{rxn} = \Delta \overline{H}^o_f(CO_2) + 2\Delta \overline{H}^o_f(H_2O) - \Delta \overline{H}^o_f(CH_4) - 2\Delta \overline{H}^o_f(O_2)$$
$$= -393.5 + 2 \times (-285.8) - (-74.4) - 2 \times 0 = -890.7 \text{ kJ}$$

b. $$\Delta H^o_{rxn} = 2 \times (-393.5) + 3 \times (-285.8) - (-83.8) - 3.5 \times 0 = -1560.6 \text{ kJ}$$

c. $$\Delta H^o_{rxn} = 2 \times (-19.5) - 2 \times (82.1) - 3 \times 0 = -203.2 \text{ kJ}$$

d. $$\Delta H^o_{rxn} = -910.7 + 2 \times (-285.8) + 2 \times 0 - 34.3 - 2 \times (-187.8) = -1141.0 \text{ kJ}$$

5. We apply [6-43] to find $\Delta H^o_{rxn,298}$:

a. $$\Delta H^o_{rxn} = 4 \times (-285.8) + 2 \times (-393.5) - 2 \times (-239.1) - 3 \times 0 = -1452.0 \text{ kJ}$$

b. $$\Delta H^o_{rxn} = 52.5 + (-83.8) - 2 \times (-74.4) - 228.2 = -110.7 \text{ kJ}$$

6. At T = 300 K, the values of $\Delta H^o_{rxn,298}$ and $\Delta S^o_{rxn,298}$ can be found from the enthalpies and entropies of formation in Table 6.1:

$$\Delta H^o_{rxn,298} = 1.9 - 0 = 1.9 \text{ kJ mol}^{-1}$$

$$\Delta S^o_{rxn,298} = 2.4 - 5.2 = -3.3 \text{ J mol}^{-1} \text{ K}^{-1}$$

$$\ln K = -\frac{1900 \text{ J mol}^{-1}}{\left(8.314 \text{ J mol}^{-1} \text{ K}^{-1}\right)\left(300 \text{ K}\right)} + \frac{-3.3 \text{ J mol}^{-1} \text{ K}^{-1}}{8.314 \text{ J mol}^{-1} \text{ K}^{-1}} = -1.158$$

$$K = 0.3137.$$

Now, we calculate ln K when T = 10^5 K, assuming that ΔH° and ΔS° remain constant.

$$\ln K = -\frac{1900 \text{ J mol}^{-1}}{\left(8.314 \text{ J mol}^{-1} \text{ K}^{-1}\right)\left(10^5 \text{ K}\right)} + \frac{-3.3 \text{ J mol}^{-1} \text{ K}^{-1}}{8.314 \text{ J mol}^{-1} \text{ K}^{-1}} = -0.399$$

$$K = 0.671.$$

We can examine how good an approximation it is to assume that ΔH°_{rxn} and ΔS°_{rxn} remain constant. The temperature dependence of ΔH°_{rxn} can be found with [6-45].

$$\Delta H^\circ_{rxn}(10^5 \text{ K}) = \Delta H^\circ_{rxn}(298) + \int_{298}^{10^5} \left(6.1 - 8.5 \text{ J K}^{-1}\right) dT$$

$$= 1900 \text{ J} + \left(-2.4 \text{ J K}^{-1}\right)\left(10^5 - 298\right) = -2.373 \times 10^5 \text{ J}.$$

To find the temperature dependence of ΔS°_{rxn}, we must recall the expression for the differential of S:

$$dS = \delta q_{rev} / T,$$

[3-17], and that at constant pressure,

$$\delta q_{rev} = C_P \, dT.$$

Combining these two relationships, we can find ΔS°_{rxn} at any temperature:

$$\Delta S^\circ_{rxn}(T) = \Delta S^\circ_{rxn}(298) + \int_{298}^{T} \frac{\Delta C_P}{T} dT.$$

$$\Delta S^\circ_{rxn}(10^5 \text{ K}) = -3.3 \text{ J mol}^{-1} \text{ K}^{-1} + \int_{298}^{10^5} \frac{-2.4 \text{ J mol}^{-1} \text{ K}^{-1}}{T} dT$$

$$= -3.3 - 2.4 \ln \frac{10^5}{298} = -17.258.$$

$$\ln K = -\frac{-2.373 \times 10^5 \text{ J mol}^{-1}}{\left(8.314 \text{ J mol}^{-1} \text{ K}^{-1}\right)\left(10^5 \text{ K}\right)} + \frac{-17.258 \text{ J mol}^{-1} \text{ K}^{-1}}{8.314 \text{ J mol}^{-1} \text{ K}^{-1}} = -2.362$$

$K = 0.0942.$

Neglecting the temperature dependence of that ΔH^{o}_{rxn} and ΔS^{o}_{rxn} leads to overstating K by a factor of 6.5.

7. The lead will be oxidized by the metals with *greater* reduction potentials, i.e., the metals below it in Table 6.2 (Cu, Hg, and Ag). To see that this is so, we reverse the Pb half-reaction so that it is written as an oxidation reaction, and its standard potential now becomes +0.1262. When coupled with any reduction half-reaction whose standard potential is greater than –0.1262, the total potential will be positive, which indicates that the overall reaction occurs spontaneously.

8. We can write the ionization reaction of sodium metal and chlorine gas as the sum of the two half-reactions:

$$Na\ (s) \rightarrow Na^+\ (aq) + e^-\ (aq) \quad E^\circ_1 = 2.71 \text{ V} \quad \Delta G^\circ_1 = -(1)\mathcal{F}\,(2.71)$$

$$Cl_2\ (g) + 2e^- \rightarrow 2Cl^-\ (aq) \quad E^\circ_2 = 1.358 \text{ V} \quad \Delta G^\circ_2 = -(2)\mathcal{F}\,(1.358)$$

The net equation we seek,

$$Na\ (s) + 0.5\ Cl_2\ (g) \rightarrow Na^+\ (aq) + Cl^-\ (aq)$$

is the sum of the first half-reaction, and the second scaled by 1/2. Therefore, the overall ΔG° is

$$\Delta G^\circ_{rxn} = \Delta G^\circ_1 + (1/2)\Delta G^\circ_2$$

$$= -(1)(96485 \text{ C mol}^{-1})(2.71) - [(2)(96485 \text{ C mol}^{-1})(1.358)]/2$$

$$= -392{,}500 \text{ C V mol}^{-1} = -392.5 \text{ kJ mol}^{-1}.$$

Additional Exercises

---

9. For a three-body collision between A, B, and C to occur, all three must occupy a particular element of volume V' within a time interval Δt. The probability that this will occur will be proportional to the probability that A is in V', B is in V', and C is in V' during Δt.

   The probability that a particular particle is in V' during the time interval Δt is equal to the ratio of the volume occupied by the particle during Δt and the total volume,

   $$P_A = \sigma_A \langle s \rangle \, \Delta t \, / \, V.$$

   The probability that any A particle occupies V' during Δt is P_A times the total number of A particles, which is the density times the volume.

   $$\mathcal{P}_A = P_A \, D \, V = (\sigma_A \langle s \rangle \, \Delta t \, / \, V) \, DV = \sigma_A \langle s \rangle \, \Delta t \;\; D.$$

   Thus, the probability that an A particle occupies V' during Δt is proportional to the average speed of A and the density. The probability that B and C will occupy V' during Δt is given by $\mathcal{P}_B$ and $\mathcal{P}_C$, which will also depend on their average speed and density. In Chapter 2, we saw that the average speed depends on the square root of the temperature [2-17].

   The probability that an A, B, and C particle will occupy V' during Δt is just the product of $\mathcal{P}_A$, $\mathcal{P}_B$, and $\mathcal{P}_C$; thus, the probability of a three-body collision will be proportional to the concentration of each species times $T^{3/2}$, i.e., $Z_{ABC} \sim [A]\,[B]\,[C]T^{3/2}$.

10. $\text{Rate} = -\dfrac{d[A]}{dt} = k[A]^3$

    $$-\int_{[A]_0}^{[A]_{t'}} \frac{1}{[A]^3} \, d[A] = \int_0^{t'} k \, dt$$

    $$\frac{1}{2\,[A]_{t'}^2} - \frac{1}{2\,[A]_0^2} = k\,t'$$

    $$[A]_{t'}^2 = \frac{[A]_0^2}{1 + 2\,k\,t\,[A]_0^2}.$$

11. From [6-33],

$$[B]_t = \frac{k_1 [A]_0}{k_2 - k_1}\left(e^{-k_1 t} - e^{-k_2 t}\right)$$

$$\frac{d[B]_t}{dt} = -\frac{k_1 [A]_0}{k_2 - k_1}\left(k_1 e^{-k_1 t} - k_2 e^{-k_2 t}\right)$$

which is the left-hand side of [6-32].

Now, we write out the right-hand side of [6-32]:

$$k_1[A]_0\, e^{-k_1 t} - k_2[B] = k_1[A]_0\, e^{-k_1 t} - \frac{k_1 k_2 [A]_0}{k_2 - k_1}\left(e^{-k_1 t} - e^{-k_2 t}\right)$$

$$= \frac{k_1 k_2 [A]_0\, e^{-k_1 t} - k_1^2 [A]_0\, e^{-k_1 t} - k_1 k_2 [A]_0\, e^{-k_1 t} + k_1 k_2 [A]_0\, e^{-k_2 t}}{k_2 - k_1}$$

$$= \frac{-k_1^2 [A]_0\, e^{-k_1 t} + k_1 k_2 [A]_0\, e^{-k_2 t}}{k_2 - k_1} = -\frac{k_1 [A]_0}{k_2 - k_1}\left(k_1 e^{-k_1 t} - k_2\, e^{-k_2 t}\right)$$

Thus, we see that [6-33] is a solution to [6-32].

12. To find the half-life for the reaction $3A \rightarrow P$, we substitute $[A]_0/2$ for $[A]_t$, and solve the rate equation from Problem 10 for $t_{1/2}$:

$$[A]_{t'}^2 = \left(\frac{[A]_0}{2}\right)^2 = \frac{[A]_0^2}{4} = \frac{[A]_0^2}{1 + 2\,k\,t_{1/2}\,[A]_0^2}.$$

$$[A]_0^2 + 2\,k\,t_{1/2}\,[A]_0^4 = 4\,[A]_0^2$$

$$2\,k\,t_{1/2}\,[A]_0^4 = 3\,[A]_0^2.$$

$$t_{1/2} = \frac{3}{2\,k\,[A]_0^2}.$$

13. To find the half-life, we must first find the integrated rate law.

$$\text{Rate} = -\frac{d[A]}{dt} = k[A]^n.$$

$$-\int_{[A]_o}^{[A]_{t'}} \frac{d[A]}{[A]^n} = \int_0^{t'} k\, dt.$$

$$\frac{1}{(n-1)\,[A]_{t'}^{n-1}} - \frac{1}{(n-1)\,[A]_0^{n-1}} = k\, t'.$$

Now, we substitute $[A]_{t'} = [A]_0/2$ to find $t_{1/2}$:

$$\frac{2^{n-1}}{(n-1)\,[A]_0^{n-1}} - \frac{1}{(n-1)\,[A]_0^{n-1}} = k\, t_{1/2}.$$

$$\frac{2^{n-1} - 1}{(n-1)k\,[A]_0^{n-1}} = t_{1/2}.$$

We can check to see if the result of the previous problem is consistent with this formula. It is.

14. $\text{Rate} = -\frac{d\,[A]}{dt} = k[A]^2[B]_0$

$$-\int_{[A]_o}^{[A]_{t'}} \frac{1}{[A]^2}\, d[A] = \int_0^{t'} k[B]_0\, dt$$

$$\frac{1}{[A]_{t'}} - \frac{1}{[A]_0} = k[B]_0\, t'$$

$$[A]_{t'} = \frac{[A]_0}{1 + k\,[B]_0\, t\,[A]_0}.$$

Notice that this equation is identical to [6-24], except that this reaction has an effective rate constant $k' = k\,[B]_0$.

15. $\text{Rate} = -\frac{d\,[A]}{dt} = k[A]^{3/2}$

$$-\int_{[A]_o}^{[A]_{t'}} \frac{1}{[A]^{3/2}}\, d[A] = \int_0^{t'} k\, dt$$

$$\frac{2}{[A]_{t'}^{1/2}} - \frac{2}{[A]_0^{1/2}} = k\,t'$$

$$[A]_{t'}^{1/2} = \frac{2[A]_0^{1/2}}{2 + k\,t\,[A]_0^{1/2}}$$

16. We label the reactions as 1 and 2, with rate constants k_1 and k_2.

$$\text{Rate} = -\frac{d\,[A]}{dt} = k_1[A]^2 - k_2[A]\,[A_2].$$

17. In Chapter 2, Problem 8, an expression was derived for the fraction of molecules with a speed greater than the most probable speed. If instead of the most probable speed, an arbitrary speed, s, was inserted as the lower limit of the integration, the fraction of molecules with speeds greater than or equal to s would be given by

$$f = \frac{2}{\sqrt{\pi}}\sqrt{\frac{m}{2kT}}\,s\,e^{-ms^2/2kT} + 1 - \text{erf}\left(\sqrt{\frac{m}{2kT}}\,s\right).$$

Here, the limit is given in terms of the kinetic energy. The speed of that kinetic energy is given by

$$s = \sqrt{\frac{(2)(\text{K.E.})}{m}}\,.$$

We can rewrite the expression for the fraction of molecules in terms of

$$\sqrt{\frac{m}{2kT}}\,s = \sqrt{\frac{m}{2kT}}\sqrt{\frac{(2)(\text{K.E.})}{m}} = \sqrt{\frac{\text{K.E.}}{kT}}$$

$$f = \frac{2}{\sqrt{\pi}}\sqrt{\frac{\text{K.E.}}{kT}}\,e^{-\text{K.E.}/kT} + 1 - \text{erf}\left(\sqrt{\frac{\text{K.E.}}{kT}}\right).$$

The kinetic energy is 100 kJ mol^{-1}, and the appropriate value for the Boltzmann constant is $k = 8.314 \times 10^{-3}$ kJ mol^{-1}.

At T = 10 K, $f = 10^{-521}$, which is essentially zero.

At T = 300 K, $f = 1.68 \times 10^{-17}$, which indicates that at molar concentrations (where the number of particles is on the order of 10^{23} L^{-1}) there would be on the order of 10^6 molecules with energy sufficient to overcome the reaction barrier.

At T = 600 K, $f = 5.86 \times 10^{-9}$. Here, there would be on the order of 10^{14} molecules with kinetic energies greater than the barrier.

Note that the mass of the particle does not explicitly enter the formula (although the kinetic energy depends on the mass). Therefore, the fractions for H_2 will be identical to those calculated here for F_2.

18. a. $$-\frac{d[S]}{dt} = k[E][S].$$

b. $$-\frac{d[S]}{dt} = k_f[E][S] - k_r[E][P].$$

c. $$-\frac{d[S]}{dt} = k_1[E][S] - k_{-1}[I]$$

$$-\frac{d[I]}{dt} = (k_{-1} + k_2)[I] - k_1[E][S] - k_{-2}[E][P].$$

In general, there is not a series of experiments that can *always* distinguish among these mechanisms. However, (a) is distinct from (b) and (c), since for (a) as $t \to \infty$, $[S] \to 0$, whereas for (b) and (c) $[S] \to [S]_\infty$. However, in comparing (b) and (c), the relative values of k_f and k_r compared to k_1, k_{-1}, k_2, and k_{-2} may be such that the two mechanisms are indistinguishable (e.g., when k_{-1} and $k_2 \ll k_1$ and k_{-2}).

19. Of the 21 gases listed in Table 6.1, the heat capacities range from 20.8 to 53.4, with an average value of 35.0. Thus, the range is nearly as large as the average value, so that using the average value to approximate the heat capacity of a real gas would be a coarse approximation. For an ideal gas, $C_P = 5/2\ R = 20.8$ J mol^{-1} K^{-1}, which corresponds to the minimum values

found (for He and Ar). The average value of C_P is greater than that of an ideal gas because of the added contribution to C_P of the vibrational degrees of freedom.

In contrast, for the 10 metals listed in the table, the heat capacities range only from 23.4 to 25.95, with an average value of 24.85. The range here is only about 10% of the average value; thus the average heat capacity is a reasonably good approximation for the heat capacity of the real metals. The picture of an ideal metal is that each atom is vibrating in three dimensions — thus, there are 3N vibrational modes — and that each one should contribute a value of k to the heat capacity according to the equipartition of energy principle (this will be explained in Chapter 11), for a total of 3Nk = 3R per mole. 3R = 24.94, which is very close to the values listed in the tables, indicating that these metals typically do behave ideally.

20. If we assume that the heat capacities are constant, [6-45] reduces to

$$\Delta H^o_{rxn}(T) = \Delta H^o_{rxn}(298) + \Delta C_{P,rxn}(T-298),$$

where $\Delta C_{P,rxn}$ is the difference in the heat capacities of the products and the reactants.

$$\Delta C_{P,rxn} = 43.6 - 43.9 - 28.8 = -29.1 \text{ J mol}^{-1} \text{ K}^{-1}.$$

$$\Delta H^o_{rxn}(298) = 52.5 - 228.2 - 0 = -175.7 \text{ kJ mol}^{-1}.$$

$$\begin{aligned}\Delta H^o_{rxn}(200) &= -175.7 \text{ kJ mol}^{-1} + (-29.1 \text{ J mol}^{-1} \text{ K}^{-1})(200-298) \\ &= -175.7 \text{ kJ mol}^{-1} + (2851.8 \text{ J mol}^{-1}) / (1000 \text{ J kJ}^{-1}) \\ &= -172.8 \text{ kJ mol}^{-1}.\end{aligned}$$

$$\begin{aligned}\Delta H^o_{rxn}(400) &= -175.7 \text{ kJ mol}^{-1} + (-29.1 \text{ J mol}^{-1} \text{ K}^{-1})(400-298) \\ &= -175.7 \text{ kJ mol}^{-1} + (-2968.2 \text{ J mol}^{-1}) / (1000 \text{ J kJ}^{-1}) \\ &= -178.7 \text{ kJ mol}^{-1}.\end{aligned}$$

$$\begin{aligned}\Delta H^o_{rxn}(800) &= -175.7 \text{ kJ mol}^{-1} + (-29.1 \text{ J mol}^{-1} \text{ K}^{-1})(800-298) \\ &= -175.7 \text{ kJ mol}^{-1} + (-14608 \text{ J mol}^{-1}) / (1000 \text{ J kJ}^{-1})\end{aligned}$$

$$= -190.3 \text{ kJ mol}^{-1}.$$

As in Problem 6, the standard entropy changes can be calculated at each temperature to then find the standard free energy changes. As shown in Problem 6,

$$\Delta S^{o}_{rxn}(T) = \Delta S^{o}_{rxn}(298) + \Delta C_{P,rxn} \ln \frac{T}{298}.$$

$$\Delta S^{o}_{rxn}(298) = 219.6 - 200.9 - 130.7 = -112.0 \text{ J mol}^{-1} \text{ K}^{-1}.$$

$$\Delta S^{o}_{rxn}(200) = -112.0 \text{ J mol}^{-1} \text{ K}^{-1} + -29.1 \text{ J mol}^{-1} \text{ K}^{-1} \ln \frac{200}{298}$$

$$= -112.0 + 11.6 = -100.4 \text{ J mol}^{-1} \text{ K}^{-1}.$$

$$\Delta S^{o}_{rxn}(400) = -112.0 \text{ J mol}^{-1} \text{ K}^{-1} + -29.1 \text{ J mol}^{-1} \text{ K}^{-1} \ln \frac{400}{298}$$

$$= -112.0 - 8.6 = -120.6 \text{ J mol}^{-1} \text{ K}^{-1}.$$

$$\Delta S^{o}_{rxn}(800) = -112.0 \text{ J mol}^{-1} \text{ K}^{-1} + -29.1 \text{ J mol}^{-1} \text{ K}^{-1} \ln \frac{800}{298}$$

$$= -112.0 - 28.7 = -140.7 \text{ J mol}^{-1} \text{ K}^{-1}.$$

$$\Delta G^{o}_{rxn}(200) = \Delta H^{o}_{rxn}(200) - T\,\Delta S_{rxn}(200)$$

$$= -172.8 - (200)(-100.4)/1000 \text{ J kJ}^{-1}$$

$$= -152.7 \text{ kJ mol}^{-1}.$$

$$\Delta G^{o}_{rxn}(400) = \Delta H^{o}_{rxn}(400) - T\,\Delta S_{rxn}(400)$$

$$= -178.7 - (400)(-120.6)/1000 \text{ J kJ}^{-1}$$

$$= -130.5 \text{ kJ mol}^{-1}.$$

$$\Delta G^{o}_{rxn}(800) = \Delta H^{o}_{rxn}(800) - T\,\Delta S_{rxn}(800)$$

$$= -190.3 - (800)(-140.7)/1000 \text{ J kJ}^{-1}$$

$$= -77.7 \text{ kJ mol}^{-1}.$$

21. For $CO + 1/2\ O_2 \rightarrow CO_2$,

$$\Delta H_{rxn,298} = -393.5 - (-110.5 + 0) = -283.0 \text{ kJ}.$$

$$\Delta S_{rxn,298} = 213.8 - (197.7 + (205.2/2)) = -86.5 \text{ J K}^{-1}.$$

$$\Delta C_{P\ rxn,298} = 37.1 - (29.1 + (29.4/2)) = -6.7 \text{ J K}^{-1}.$$

As in Problem 6, we can now calculate ΔH and ΔS at $T = 1000$ K:

$$\Delta H_{rxn,1000} = \Delta H_{rxn,298} + \Delta C_{P\ rxn,298}\,(1000 - 298)$$

$$= (-283.0 \text{ kJ} \times 1000 \text{ J kJ}^{-1}) + (-6.7 \text{ J K}^{-1})(702 \text{ K})$$

$$= -287700 \text{ J}.$$

$$\Delta S_{rxn,1000} = \Delta S_{rxn,298} + \Delta C_{P\ rxn,298} \ln (1000/298)$$

$$= -86.5 \text{ J K}^{-1} + (-6.7 \text{ J K}^{-1}) \ln (1000/298)$$

$$= -94.6 \text{ J K}^{-1}.$$

$$\ln K = -(-287700 \text{ J}) / (8.314 \text{ J K}^{-1} \times 1000 \text{ K}) + (-94.6 \text{ J K}^{-1}/8.314 \text{ J K}^{-1})$$

$$= 23.23.$$

$$K = 1.22 \times 10^{10}.$$

To go beyond approximating the heat capacities as constant, we can approximate them as a linear function between 298 and 1200 K:

$$\Delta C_{P,298} = 37.135 - (29.141 + (29.378/2)) = -6.695 \text{ J K}^{-1}.$$

$$\Delta C_{P,1200} = 56.354 - (34.169 + (35.683/2)) = 4.344 \text{ J K}^{-1}.$$

$$\Delta C_P (T) = -6.695 + [(4.344 - (-6.695))/(1200 - 298)]\, T$$

$$= -6.695 + 0.01224\, T \text{ (K)}.$$

Now, we use this function to represent the heat capacity as a function of temperature:

$$\Delta H_{rxn,1000} = -283.0 \text{ kJ} + \int_{298}^{1000} (-6.695 + 0.01224 \text{ T}) \text{ dT}$$

$$= -283.0 + (-6.695) \times (1000 - 298) + \frac{0.01224}{2}(1000^2 - 298^2)$$

$$= -283{,}000 \text{ J} + 876.6 \text{ J} = -282{,}123 \text{ J}.$$

$$\Delta S_{rxn,1000} = -86.5 \text{ J K}^{-1} + \int_{298}^{1000} \left(\frac{-6.695 + 0.01224 \text{ T}}{\text{T}} \right) \text{dT}$$

$$= -86.5 + (-6.695) \times \ln(1000 / 298) + 0.01224 \times (1000 - 298)$$

$$= -86.5 \text{ J} + 0.487 \text{ J} = -86.013 \text{ J}.$$

$$\ln K = -(-282123 \text{ J}) / (8.314 \text{ J K}^{-1} \times 1000 \text{ K}) + (-86.013 \text{ J K}^{-1} / 8.314 \text{ J K}^{-1})$$

$$= 23.59.$$

$$K = 1.75 \times 10^{10}.$$

The effect of the temperature dependence of the heat capacity leads to only a 50% increase in the equilibrium constant.

22. From the previous problem,

$$\Delta H_{rxn,600} = -283.0 \text{ kJ} + \int_{298}^{600} (-6.695 + 0.01224 \text{ T}) \text{ dT}$$

$$= -283.0 + (-6.695) \times (600 - 298) + \frac{0.01224}{2}(600^2 - 298^2)$$

$$= -283{,}000 \text{ J} + -362.17 \text{ J} = -283362 \text{ J}.$$

$$\Delta S_{rxn,600} = -86.5 \text{ J K}^{-1} + \int_{298}^{600} \left(\frac{-6.695 + 0.01224 \text{ T}}{\text{T}} \right) \text{dT}$$

$$= -86.5 + (-6.695) \times \ln(600 / 298) + 0.01224 \times (600 - 298)$$

$$= -86.5 \text{ J} + -0.989 \text{ J} = -87.489 \text{ J}.$$

$$\Delta G_{rxn,600} = \Delta H_{rxn,600} - T \, \Delta S_{rxn,600} = -283{,}362 - (600)(-87.489)$$

$= -230{,}869 \text{ J} = -230.9 \text{ kJ}.$

23. We consider the reaction of the borate ion and bromide ion in acidic solution:

$$BrO_3^- (aq) + 5\ Br^- (aq) + 6\ H^+ \longrightarrow 3\ Br_2\ (g) + 3\ H_2O\ (l)$$

The progress of this reaction could be monitored in terms of the concentration of the H^+ ions. The $[H^+]$ can be measured with an electrochemical cell.

The experimental setup would consist of two gas cells, the potentiometer, and the timer. The gas cells would be like those pictured in Figure 6.10, only in this case both would be hydrogen cells, with the H_2 gas at a pressure of 1 bar. In one cell, the $[H^+]$ would be held at 1.0 M; in the other cell all the reactants (BrO_3^-, Br^-, and H^+) would be introduced at the starting time. The potentiometer would connect the two cells and would be adjusted to generate a potential just the opposite of the cell, so that no current flows. This potential would depend solely on the $[H^+]$.

Reaction cell: $H^+ (aq) + e^- \longrightarrow H_2 (g)$ $\quad E° = 0.00$

Standard cell: $H_2 (g) \longrightarrow H^+ (aq) + e^-$ $\quad E° = 0.00$

Nernst equation: $E = E° - (RT/n\mathcal{F}) \ln Q$

$$E = (RT/n\mathcal{F}) \ln [H^+]_{react}$$

This setup allows for the measurement of the $[H^+]$ as a function of time, by which the reaction kinetics can be studied.

24. a. The two half-reactions that form this redox reaction are

$Mg (s) \rightarrow Mg^{2+} + 2\ e^-$ $\quad E° = 2.372$ V

$Cl_2 (g) + 2\ e^- \rightarrow 2\ Cl^-$ $\quad E° = 1.358$ V

$Mg (s) + Cl_2 (g) \rightarrow MgCl_2$ $\quad E° = 3.730$ V

$\Delta G° = -2 \text{ mol } e^- \times 96485 \text{ C mol}^{-1} \times 3.730 \text{ V} = -719778 \text{ J} = -719.8 \text{ kJ}.$

The activity quotient for this reaction is

$$Q = [MgCl_2] / P_{Cl_2}.$$

Therefore, doubling the pressure will reduce Q, which will decrease the cell voltage, since $E = E° - (RT/n\mathcal{F}) \ln Q$ [6-66].

b. For this reaction,

$$Li\,(s) \rightarrow Li^+ + e^- \qquad E° = 3.0401\ V$$

$$K^+ + e^- \rightarrow K\,(s) \qquad E° = -2.931\ V$$

$$Li\,(s) + K^+ \rightarrow Li^+ + K\,(s) \qquad E° = 0.1091\ V$$

$\Delta G° = -\,(1\ mol\ e^-)\,(96485\ C\ mol^{-1})\,(0.1091\ V) = -10526\ J = -10.53\ kJ.$

Because none of the reactants are in the gas phase, the voltage is independent of pressure.

25. The two half-reactions that form this redox reaction are

$$Pb\,(s) \rightarrow Pb^{2+} + 2\,e^- \qquad E° = 0.1262\ V$$

$$Cl_2\,(g) + 2\,e^- \rightarrow 2\,Cl^- \qquad E° = 1.358\ V$$

$$Pb\,(s) + Cl_2\,(g) \rightarrow PbCl_2 \qquad E° = 1.4842\ V$$

$\Delta G° = -2\ mol\ e^- \times 96485\ C\ mol^{-1} \times 1.4842\ V = -286406\ J = -286.4\ kJ.$

The activity quotient for this reaction is

$$Q = [PbCl_2] / P_{Cl_2}.$$

Therefore, doubling the pressure will reduce Q, which will decrease the cell voltage, since $E = E° - (RT/n\mathcal{F}) \ln Q$ [6-66].

Chapter 7

Vibrational Mechanics of Particle Systems

Exercises

1. $x(t) = x_o + a\sin(\omega t) + b\cos(\omega t)$ [7-12a]

$$\frac{d\,x(t)}{dt} = \dot{x}(t) = a\omega\cos(\omega t) - b\omega\sin(\omega t)$$

$$\frac{d\,\dot{x}(t)}{dt} = \ddot{x}(t) = -a\omega^2\sin(\omega t) - b\omega^2\cos(\omega t)$$

$$\ddot{x}(t) = -\omega^2[a\sin(\omega t) + b\cos(\omega t)]$$

$$= -\omega^2[x_o + a\sin(\omega t) + b\cos(\omega t) - x_o]$$

$$= -\omega^2[x(t) - x_o].$$

$x(t) = x_o + A\,e^{(-i\omega t)} + B\,e^{(i\omega t)}$ [7-12b]

$$\frac{d\,x(t)}{dt} = \dot{x}(t) = -i\omega A\,e^{-i\omega t} + i\omega B\,e^{i\omega t}.$$

$$\frac{d\,\dot{x}(t)}{dt} = \ddot{x}(t) = (-i\omega)^2\, A\, e^{-i\omega t} + (i\omega)^2\, B\, e^{i\omega t}$$

$$= -\omega^2[A\, e^{-i\omega t} + B\, e^{i\omega t}]$$

$$= -\omega^2[x_o + A\, e^{-i\omega t} + B\, e^{i\omega t} - x_o]$$

$$= -\omega^2[x(t) - x_o].$$

2. $k = \omega^2 m = (10\ s^{-1})^2\,(1.0\ kg) = 100\ kg\ s^{-2} = 100\ N\ m^{-1}$.

a. $\omega = (k/m)^{1/2} = [\,(100\ N\ m^{-1})\,(9.109 \times 10^{-31}\ kg)\,]^{1/2} = 1.048 \times 10^{16}\ s^{-1}$.

b. hydrogen: $1.0078\ amu \times (1.661 \times 10^{-27}\ kg\ amu^{-1}) = 1.673 \times 10^{-27}\ kg$.

$\omega = (k/m)^{1/2} = [\,(100\ N\ m^{-1})\,(1.673 \times 10^{-27}\ kg)\,]^{1/2} = 2.445 \times 10^{14}\ s^{-1}$.

c. $\omega = (k/m)^{1/2} = [\,(100\ N\ m^{-1})\,(0.005\ kg)\,]^{1/2} = 1.414 \times 10^{2}\ s^{-1}$.

3. $$\lambda = \frac{h}{p} = \frac{6.626 \times 10^{-34}\ J\ s}{(9.109 \times 10^{-31}\ kg)\,(0.1 \times 2.998 \times 10^{8}\ m\ s^{-1})}$$

$$= 2.426 \times 10^{-11}\ m.$$

$$s = \frac{h}{m\lambda} = \frac{6.626 \times 10^{-34}\ J\ s}{(0.01\ kg)\,(2.426 \times 10^{-11}\ m)} = 2.731 \times 10^{-21}\ m\ s^{-1}.$$

4. The mass of He $= 6.6465 \times 10^{-27}$ kg (Prob. 1-7). From [2-17],

$$\langle s \rangle = \sqrt{\frac{8kT}{\pi m}} = \sqrt{\frac{(8)(1.381 \times 10^{-23}\ J\ K^{-1})(300\ K)}{(3.1416)(6.6465 \times 10^{-27}\ kg)}} = 1259.9\ m\ s^{-1}.$$

$$\lambda = \frac{h}{p} = \frac{6.626 \times 10^{-34}\ J\ s}{(6.6465 \times 10^{-27}\ kg)\,(1259.9\ m\ s^{-1})} = 7.913 \times 10^{-11}\ m.$$

5. Since the frequency of the oscillator is $\nu = 1$ cycle s^{-1}, the vibrational state energies are given by

$$E_n = (n + 1/2)\,h\nu = (n + 1/2)\,(6.626 \times 10^{-34}\ J\ s)\,(1.0\ cycle\ s^{-1}).$$

[Note: When the frequency is stated in cycles s^{-1}, it is designated by ν; when stated in s^{-1} or radians s^{-1}, it is designated by ω. Since $\omega = 2\pi\nu$,

$$\hbar\omega = \frac{h}{2\pi}\omega = h\frac{\omega}{2\pi} = h\nu.$$

Therefore, we can write [7-34] as $E_n = (n + 1/2)\, h\nu$ or $E_n = (n + 1/2)\, \hbar\omega$.]

$$\Delta E = (n + 1 + 1/2)\, h\nu - (n + 1/2)\, h\nu = h\nu = 6.626 \times 10^{-34} \text{ J}.$$

$$E_0 = (1/2)\,(6.626 \times 10^{-34} \text{ J}) = 3.313 \times 10^{-34} \text{ J (the zero-point energy)}.$$

These energies are so infinitesimally small compared with macroscopic scales of energy, that the allowed energies of the oscillator would appear continuous.

$$n \approx 10^{-10} \text{ J} / 10^{-34} \text{ J} \approx 10^{24}.$$

6. Hydrogen mass: 1.00783 amu × 1.661 × 10^{-27} kg amu^{-1} = 1.674 × 10^{-27} kg.
   Deuterium mass: 2.0141 amu × 1.661 × 10^{-27} kg amu^{-1} = 3.345 × 10^{-27} kg.
   ^{35}Cl mass: 34.9689 amu × 1.661 × 10^{-27} kg amu^{-1} = 5.808 × 10^{-26} kg.

$$\mu_{HCl} = \frac{(1.674 \times 10^{-27} \text{ kg})(5.808 \times 10^{-26} \text{ kg})}{(1.674 \times 10^{-27} \text{ kg} + 5.808 \times 10^{-26} \text{ kg})} = 1.627 \times 10^{-27} \text{ kg}.$$

$$\mu_{DCl} = \frac{(3.345 \times 10^{-27} \text{ kg})(5.808 \times 10^{-26} \text{ kg})}{(3.345 \times 10^{-27} \text{ kg} + 5.808 \times 10^{-26} \text{ kg})} = 3.163 \times 10^{-27} \text{ kg}.$$

$$\mu_{HCl} / \mu_{DCl} = 1.627 \times 10^{-27} \text{ kg} / 3.163 \times 10^{-27} \text{ kg} = 0.5144.$$

7. $k = (2\pi\nu)^2 m = [(2)(3.1416)(1.0 \times 10^{14} \text{ cycles } s^{-1})]^2\, (1.674 \times 10^{-27} \text{ kg})$
   $= 660.9 \text{ kg } s^{-2}$. (See note regarding ν and ω in Problem 5.)

   At the classical turning point, the potential energy ($1/2\, kx^2$) is equal to the vibrational state energy.

$$E_0 = (1/2)\, h\nu = (1/2)\,(6.626 \times 10^{-34} \text{ J s})(10^{14} \text{ s}^{-1}) = 3.313 \times 10^{-20} \text{ J}.$$

$$x_{\text{turning point}} = \sqrt{\frac{2\,E_0}{k}} = \sqrt{\frac{(2)(3.313 \times 10^{-20}\ \text{J})}{660.9\ \text{kg s}^{-2}}} = 1.001 \times 10^{-11}\ \text{m}.$$

$E_1 = (3/2)\,h\nu = (3/2)\,(6.626 \times 10^{-34}\ \text{J s})(10^{14}\ \text{s}^{-1}) = 9.939 \times 10^{-20}\ \text{J}.$

$$x_{\text{turning point}} = \sqrt{\frac{2\,E_1}{k}} = \sqrt{\frac{(2)(9.939 \times 10^{-20}\ \text{J})}{660.9\ \text{kg s}^{-2}}} = 1.734 \times 10^{-11}\ \text{m}.$$

$E_2 = (5/2)\,h\nu = (5/2)\,(6.626 \times 10^{-34}\ \text{J s})(10^{14}\ \text{s}^{-1}) = 1.657 \times 10^{-19}\ \text{J}.$

$$x_{\text{turning point}} = \sqrt{\frac{2\,E_2}{k}} = \sqrt{\frac{(2)(1.657 \times 10^{-19}\ \text{J})}{660.9\ \text{kg s}^{-2}}} = 2.239 \times 10^{-11}\ \text{m}.$$

$E_3 = (7/2)\,h\nu = (7/2)\,(6.626 \times 10^{-34}\ \text{J s})(10^{14}\ \text{s}^{-1}) = 2.319 \times 10^{-19}\ \text{J}.$

$$x_{\text{turning point}} = \sqrt{\frac{2\,E_3}{k}} = \sqrt{\frac{(2)(2.319 \times 10^{-19}\ \text{J})}{660.9\ \text{kg s}^{-2}}} = 2.649 \times 10^{-11}\ \text{m}.$$

Additional Exercises

8. $H(x,y,z,p_x,p_y,p_z) = T + V = (p_x^2 + p_y^2 + p_z^2) / 2m + mgz$. For this particular problem, we replace the generalized position coordinates q_1, q_2, and q_3 with x, y, and z in [7-5] and [7-6], so

$$\frac{\partial H}{\partial x} = -\dot{p}_x \Rightarrow \dot{p}_x = 0;\ \frac{\partial H}{\partial y} = -\dot{p}_y \Rightarrow \dot{p}_y = 0;\frac{\partial H}{\partial z} = -\dot{p}_z \Rightarrow \dot{p}_z = mg.$$

and

$$\frac{\partial H}{\partial p_x} = \dot{x} \Rightarrow \dot{x} = \frac{p_x}{m};\frac{\partial H}{\partial p_y} = \dot{y} \Rightarrow \dot{y} = \frac{p_y}{m};\frac{\partial H}{\partial p_z} = \dot{z} \Rightarrow \dot{z} = \frac{p_z}{m}.$$

9. When the particles are both free to move in three-dimensional space, the positions of the two masses can be described by x_1, y_1, z_1, x_2, y_2 and z_2. The equilibrium position of the spring is r_0, i.e., the potential energy of the spring is at its minimum value when the separation of the two particles is r_0.

$$H = \frac{p_{x_1}{}^2}{m_1} + \frac{p_{y_1}{}^2}{m_1} + \frac{p_{z_1}{}^2}{m_1} + \frac{p_{x_2}{}^2}{m_2} + \frac{p_{y_2}{}^2}{m_2} + \frac{p_{z_2}{}^2}{m_2} +$$

$$+ \frac{1}{2}k\left\{\left[(x_2 - x_1)^2 + (y_2 - y_1)^2 + (z_2 - z_1)^2\right]^{1/2} - r_0\right\}^2$$

$$+ \frac{Q_1Q_2}{\left[(x_2 - x_1)^2 + (y_2 - y_1)^2 + (z_2 - z_1)^2\right]^{1/2}}$$

10. Using the stated definition of $e^{-i\alpha}$ and the trigonometric identities $\cos(x) = \cos(-x)$ and $\sin(x) = -\sin(-x)$, we obtain

$$A\, e^{-i\omega t} = A\cos(-\omega t) + iA\sin(-\omega t) = A\cos(\omega t) - iA\sin(\omega t).$$

$$B\, e^{i\omega t} = B\cos(\omega t) + iB\sin(\omega t).$$

$$A\, e^{-i\omega t} + B\, e^{i\omega t} = (A + B)\cos(\omega t) + (B - A)\, i\sin(\omega t).$$

For this expression to equal $a\sin(\omega t) + b\cos(\omega t)$, then

$$A + B = b \quad \text{and} \quad (B - A)i = a.$$

For $b = 0$, that implies that $A = -B$. Then, if $a = 1$,

$$1 = (B - (-B))i = 2Bi\ , \quad \text{and } B = 1/2i.\ \ A = -1/2i.$$

In general, for any value of a,

$$B = a/2i \text{ and } A = -a/2i.$$

11. We define a coordinate system where the origin is at the lower left corner of the square frame. The fixed points of the four springs attached to the frame are then $(0,0)$, $(l,0)$, $(0,l)$, and (l,l), where l is the length of a side.

Each spring will contribute one term to the total potential energy of the form given in [7-7], $1/2\ k\ (r - r_0)^2$. For each term, the extension of the spring, r, must be found from the geometry given. For example, the extension of the spring connected to m_1 with force constant k_A is

$$r = [(x_1 - 0)^2 + (y_1 - l)^2]^{1/2}$$

since the fixed point of the spring lies at $(0,l)$. Thus the potential energy for this spring is given by

$$V_{1,A} = 1/2\ k_A\ \{[(x_1 - 0)^2 + (y_1 - l)^2]^{1/2} - s_A\}^2.$$

The total potential energy is the sum of all the individual terms.

$$\begin{aligned} V = 1/2\ k_A \Big(& \{[x_1^2 + (l - y_1)^2]^{1/2} - s_A\}^2 + \{[(l - x_2)^2 + (l - y_2^2)]^{1/2} - s_A\}^2 \\ & + \{[(l - x_3)^2 + y_3^2]^{1/2} - s_A\}^2 + \{[x_4^2 + y_4^2]^{1/2} - s_A\}^2 \Big) \\ + 1/2\ k_B \Big(& \{[(x_1 - x_2)^2 + (y_1 - y_2)^2]^{1/2} - s_B\}^2 \\ & + \{[(x_2 - x_3)^2 + (y_2 - y_3)^2]^{1/2} - s_B\}^2 \\ & + \{[(x_3 - x_4)^2 + (y_3 - y_4)^2]^{1/2} - s_B\}^2 \\ & + \{[(x_4 - x_1)^2 + (y_4 - y_1)^2]^{1/2} - s_B\}^2 \Big). \end{aligned}$$

12. The kinetic energy term in the Hamiltonian is straightforward.

$$T = \sum_{i=1}^{N} \frac{p_i^2}{2m}.$$

The potential energy is the sum of the potential energy of each spring. Since there are N point masses, they are connected by N – 1 springs. If we designate spring i as the spring connecting particles i and i + 1 and $x_{0,i}$ as the equilibrium separation of spring i, we can write the potential energy as

$$V = \sum_{i=1}^{N-1} \frac{1}{2} k_i (x_{i+1} - x_i - x_{0,i})^2 .$$

The total Hamiltonian is the sum of T and V:

$$H = \sum_{i=1}^{N} \frac{p_i^{\,2}}{2m} + \sum_{i=1}^{N-1} \frac{1}{2} k_i (x_{i+1} - x_i - x_{0,i})^2 .$$

Now, we can apply Hamilton's equations to particle i to obtain the differential equations that must be solved to describe this system.

$$\frac{\partial H}{\partial p_i} = \frac{p_i}{m} = \dot{x}_i, \text{ for all } i = 1,2,3, \ldots, N.$$

$$\frac{\partial H}{\partial x_i} = k_{i-1}(x_i - x_{i-1} - x_{0,i-1}) - k_i (x_{i+1} - x_i - x_{0,i}) = -\dot{p}_i$$

for $i = 2,3, \ldots, N-1$, because the position of each particle contributes to the potential energy of the spring connecting it to the particles before and after it. The first and last particles are exceptions, and their equations are

$$\frac{\partial H}{\partial x_1} = -k_1 (x_2 - x_1 - x_{0,1}) = -\dot{p}_1 .$$

$$\frac{\partial H}{\partial x_N} = k_{N-1}(x_N - x_{N-1} - x_{0,N-1}) = -\dot{p}_N .$$

13. $$\begin{aligned} H = {} & \sum_{i=1}^{4} \left(\frac{p_{x_i}^{\,2}}{2m_i} + \frac{p_{y_i}^{\,2}}{2m_i}\right) + 1/2\, k_A \Big(\{[x_4^{\,2} + y_4^{\,2}]^{1/2} - s_A\}^2 \\ & + \{[(l - x_3)^2 + y_3^{\,2}]^{1/2} - s_A\}^2 + \{[x_1^{\,2} + (l - y_1)^2]^{1/2} - s_A\}^2 \\ & + \{[(l - x_2)^2 + (l - y_2^{\,2})]^{1/2} - s_A\}^2\Big) \\ & + 1/2\, k_B \Big(\{[(x_1 - x_2)^2 + (y_1 - y_2)^2]^{1/2} - s_B\}^2 \\ & + \{[(x_2 - x_3)^2 + (y_2 - y_3)^2]^{1/2} - s_B\}^2 \\ & + \{[(x_3 - x_4)^2 + (y_3 - y_4)^2]^{1/2} - s_B\}^2 \\ & + \{[(x_4 - x_1)^2 + (y_4 - y_1)^2]^{1/2} - s_B\}^2\Big). \end{aligned}$$

For each of the eight coordinates, there will be a pair of equations. The first of the pair comes from applying [7-6]

$$\frac{\partial H}{\partial p_{x_i}} = \frac{p_{x_i}}{m_i} = \dot{x}_i \text{ and } \frac{\partial H}{\partial p_{y_i}} = \frac{p_{y_i}}{m_i} = \dot{y}_i, \text{ i=1,2,3,4.}$$

The second of each pair comes from applying [7-5]:

$$-\dot{p}_{x_1} = \frac{\partial H}{\partial x_1} = k_A\{[x_1^2 + (l - y_1)^2]^{1/2} - s_A\}[x_1^2 + (l - y_1)^2]^{-1/2}(x_1)$$
$$+ k_B\{[(x_1 - x_2)^2 + (y_1 - y_2)^2]^{1/2} - s_B\}[(x_1 - x_2)^2 + (y_1 - y_2)^2]^{-1/2}(x_1 - x_2)$$
$$+ k_B\{[(x_4 - x_1)^2 + (y_4 - y_1)^2]^{1/2} - s_B\}[(x_4 - x_1)^2 + (y_4 - y_1)^2]^{-1/2}(x_4 - x_1)(-1).$$

$$-\dot{p}_{x_2} = \frac{\partial H}{\partial x_2} = k_A\{[(l - x_2)^2 + (l - y_2)^2]^{1/2} - s_A\}[(l - x_2)^2 + (l - y_2)^2]^{-1/2}(l - x_2)(-1)$$
$$+ k_B\{[(x_1 - x_2)^2 + (y_1 - y_2)^2]^{1/2} - s_B\}[(x_1 - x_2)^2 + (y_1 - y_2)^2]^{-1/2}(x_1 - x_2)(-1)$$
$$+ k_B\{[(x_2 - x_3)^2 + (y_2 - y_3)^2]^{1/2} - s_B\}[(x_2 - x_3)^2 + (y_2 - y_3)^2]^{-1/2}(x_2 - x_3).$$

$$-\dot{p}_{x_3} = \frac{\partial H}{\partial x_3} = k_A\{[(l - x_3)^2 + y_3^2]^{1/2} - s_A\}[(l - x_3)^2 + y_3^2]^{-1/2}(l - x_3)(-1)$$
$$+ k_B\{[(x_2 - x_3)^2 + (y_2 - y_3)^2]^{1/2} - s_B\}[(x_2 - x_3)^2 + (y_2 - y_3)^2]^{-1/2}(x_2 - x_3)(-1)$$
$$+ k_B\{[(x_3 - x_4)^2 + (y_3 - y_4)^2]^{1/2} - s_B\}[(x_3 - x_4)^2 + (y_3 - y_4)^2]^{-1/2}(x_3 - x_4).$$

$$-\dot{p}_{x_4} = \frac{\partial H}{\partial x_4} = k_A\{[x_4^2 + y_4^2]^{1/2} - s_A\}[x_4^2 + y_4^2]^{-1/2}(x_4)$$
$$+ k_B\{[(x_3 - x_4)^2 + (y_3 - y_4)^2]^{1/2} - s_B\}[(x_3 - x_4)^2 + (y_3 - y_4)^2]^{-1/2}(x_3 - x_4)(-1)$$
$$+ k_B\{[(x_4 - x_1)^2 + (y_4 - y_1)^2]^{1/2} - s_B\}[(x_4 - x_1)^2 + (y_4 - y_1)^2]^{-1/2}(x_4 - x_1).$$

$$-\dot{p}_{y_1} = \frac{\partial H}{\partial y_1} = k_A\{[x_1^2 + (l - y_1)^2]^{1/2} - s_A\}[x_1^2 + (l - y_1)^2]^{-1/2}(l - y_1)(-1)$$
$$+ k_B\{[(x_1 - x_2)^2 + (y_1 - y_2)^2]^{1/2} - s_B\}[(x_1 - x_2)^2 + (y_1 - y_2)^2]^{-1/2}(y_1 - y_2)$$
$$+ k_B\{[(x_4 - x_1)^2 + (y_4 - y_1)^2]^{1/2} - s_B\}[(x_4 - x_1)^2 + (y_4 - y_1)^2]^{-1/2}(y_4 - y_1)(-1).$$

$$-\dot{p}_{y_2} = \frac{\partial H}{\partial y_2} = k_A\{[(l - x_2)^2 + (l - y_2)^2]^{1/2} - s_A\}[(l - x_2)^2 + (l - y_2)^2]^{-1/2}(l - y_2)(-1)$$
$$+ k_B\{[(x_1 - x_2)^2 + (y_1 - y_2)^2]^{1/2} - s_B\}[(x_1 - x_2)^2 + (y_1 - y_2)^2]^{-1/2}(y_1 - y_2)(-1)$$

$$+ k_B\{[(x_2 - x_3)^2 + (y_2 - y_3)^2]^{1/2} - s_B\}[(x_2 - x_3)^2 + (y_2 - y_3)^2]^{-1/2}(y_2 - y_3).$$

$$-\dot{p}_{y_3} = \frac{\partial H}{\partial y_3} = k_A\{[(l - x_3)^2 + y_3^2]^{1/2} - s_A\}[(l - x_3)^2 + y_3^2]^{-1/2}(y_3)$$
$$+ k_B\{[(x_2 - x_3)^2 + (y_2 - y_3)^2]^{1/2} - s_B\}[(x_2 - x_3)^2 + (y_2 - y_3)^2]^{-1/2}(y_2 - y_3)(-1)$$
$$+ k_B\{[(x_3 - x_4)^2 + (y_3 - y_4)^2]^{1/2} - s_B\}[(x_3 - x_4)^2 + (y_3 - y_4)^2]^{-1/2}(y_3 - y_4).$$

$$-\dot{p}_{y_4} = \frac{\partial H}{\partial y_4} = k_A\{[x_4^2 + y_4^2]^{1/2} - s_A\}[x_4^2 + y_4^2]^{-1/2}(y_4)$$
$$+ k_B\{[(x_3 - x_4)^2 + (y_3 - y_4)^2]^{1/2} - s_B\}[(x_3 - x_4)^2 + (y_3 - y_4)^2]^{-1/2}(y_3 - y_4)(-1)$$
$$+ k_B\{[(x_4 - x_1)^2 + (y_4 - y_1)^2]^{1/2} - s_B\}[(x_4 - x_1)^2 + (y_4 - y_1)^2]^{-1/2}(y_4 - y_1).$$

14. In Cartesian coordinates, where the position of the particle is given by x, y, z, the Hamiltonian is

$$\hat{H} = \frac{p_x^2}{2m} + \frac{p_y^2}{2m} + \frac{p_z^2}{2m} + \frac{1}{2}k([x^2 + y^2 + z^2]^{1/2} - r_o)^2,$$

where the pivot is located at (0,0,0) and the equilibrium separation of the spring is r_0.

Now, we consider two particles free to move in three-dimensional space, connected by a harmonic spring. The Hamiltonian for this system was given in Problem 9 (if the electrostatic term is neglected):

$$H = \frac{p_{x_1}^2}{2m_1} + \frac{p_{y_1}^2}{2m_1} + \frac{p_{z_1}^2}{2m_1} + \frac{p_{x_2}^2}{2m_2} + \frac{p_{y_2}^2}{2m_2} + \frac{p_{z_3}^2}{2m_2} +$$
$$+ \frac{1}{2}k\left\{\left[(x_2 - x_1)^2 + (y_2 - y_1)^2 + (z_2 - z_1)^2\right]^{1/2} - r_0\right\}^2 .$$

To see how these two systems are equivalent, we make the transformation from Cartesian coordinates to the center-of-mass coordinates, X, Y, Z, and relative displacement coordinates, x,y,z. These coordinates are defined by

$$X = \frac{m_1 x_1 + m_2 x_2}{m_1 + m_2} \quad Y = \frac{m_1 y_1 + m_2 y_2}{m_1 + m_2} \quad Z = \frac{m_1 z_1 + m_2 z_2}{m_1 + m_2}.$$

$$x = x_2 - x_1 \qquad y = y_2 - y_1 \qquad z = z_2 - z_1.$$

Following the example in 7.2, we rewrite the Cartesian coordinates in terms of the center-of-mass and displacement coordinates, and substitute the result back into the Hamiltonian.

$$X + \frac{m_1}{m_1 + m_2} x = \frac{m_1 x_1 + m_2 x_2}{m_1 + m_2} + \frac{m_1 x_2 - m_1 x_1}{m_1 + m_2} = \frac{m_1 + m_2}{m_1 + m_2} x_2 = x_2.$$

$$X - \frac{m_2}{m_1 + m_2} x = \frac{m_1 x_1 + m_2 x_2}{m_1 + m_2} - \frac{m_2 x_2 - m_2 x_1}{m_1 + m_2} = \frac{m_1 + m_2}{m_1 + m_2} x_1 = x_1.$$

These relationships apply to the y and z coordinates as well:

$$x_2 = X + \frac{m_1}{m_1 + m_2} x, \quad y_2 = Y + \frac{m_1}{m_1 + m_2} y, \quad z_2 = Z + \frac{m_1}{m_1 + m_2} z.$$

$$x_1 = X - \frac{m_2}{m_1 + m_2} x, \quad y_1 = Y - \frac{m_2}{m_1 + m_2} y, \quad z_1 = Z + \frac{m_2}{m_1 + m_2} z.$$

For each coordinate, the associated momentum is formed by multiplying the time derivative by the mass of the particle.

$$p_{x_1} = m_1 \dot{x}_1 = m_1 \left(\dot{X} - \frac{m_2}{m_1 + m_2} \dot{x} \right), \quad p_{x_2} = m_2 \dot{x}_2 = m_2 \left(\dot{X} + \frac{m_1}{m_1 + m_2} \dot{x} \right)$$

By squaring both of the x-direction momenta, we get

$$\frac{p_{x_1}{}^2}{2m_1} + \frac{p_{x_2}{}^2}{2m_2} = \frac{p_X{}^2}{2(m_1 + m_2)} + \frac{p_x{}^2}{2\left(\dfrac{m_1 m_2}{m_1 + m_2} \right)}$$

Let $M = m_1 + m_2$ (total mass) and $\mu = m_1 m_2 / (m_1 + m_2)$ (the reduced mass); we can now write the Hamiltonian as

$$H = \frac{p_X{}^2}{2M} + \frac{p_Y{}^2}{2M} + \frac{p_Z{}^2}{2M} + \frac{p_x{}^2}{2\mu} + \frac{p_y{}^2}{2\mu} + \frac{p_z{}^2}{2\mu} +$$

$$+ \frac{1}{2}k\left[\left(x^2 + y^2 + z^2\right)^{1/2} - r_0\right]^2$$

This Hamiltonian is identical with that written for the mass attached to the pivot, except that this system has the additional translational motion of the entire assembly (i.e., motion of the center of mass described by X, Y, Z).

15. If all four masses are equal, with the given spring constants, the symmetric stretch will retain the square symmetry of the box-and-spring arrangement. Each atom will move toward the corners so that the masses remain in a square as they move closer to or farther from the center in unison.

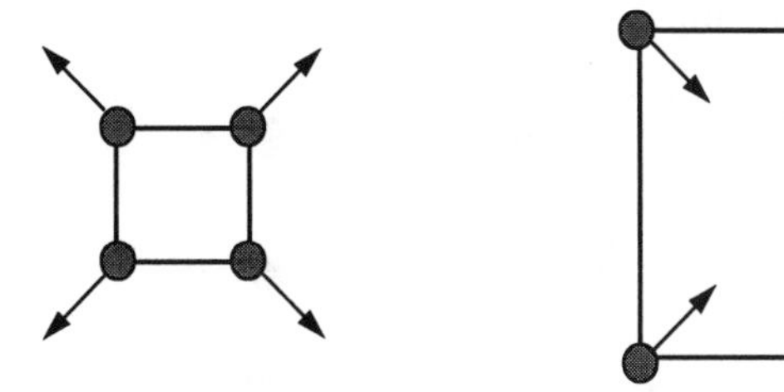

If m_1 and m_3, the masses on opposite corners, are twice as heavy as m_2 and m_4, the masses on the other set of opposite corners, the square symmetry will be broken. The heavier masses will move more slowly, leading to diamond shaped conformations.

Without doing so explicitly, we based this analysis on the symmetry of the Hamiltonian operator. If all four masses are of identical mass, then the Hamiltonian remains unchanged under a C_4 symmetry operation (see Appendix II: Molecular Symmetry for a more detailed discussion). When only the masses on opposite corners are identical, the lower symmetry C_2 operator applies.

16. The number of normal modes is equal to the total number of degrees of freedom, so for these linear systems, the number of normal modes is the same as the number of particles. For three particles, there must be three normal modes. One mode represents the movement of all three particles in unison, so that only the springs attached to the walls are vibrating; the internal springs remain at a constant extension. In the symmetric stretch for three particles the outer particles move in equal and opposite directions while the middle one remains fixed. The third mode is an asymmetric stretch in which the outer particles move in the same direction and the middle moves in opposition.

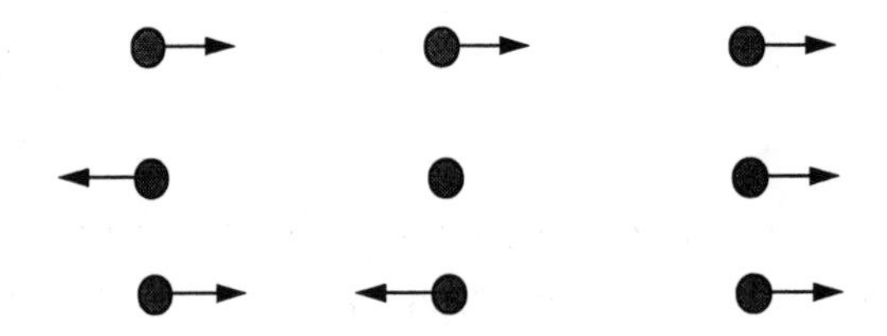

As was discussed in the text, the first normal mode has a smaller frequency, since in this mode only two of the four springs are out of equilibrium.

For the system of four particles, there are four normal modes:

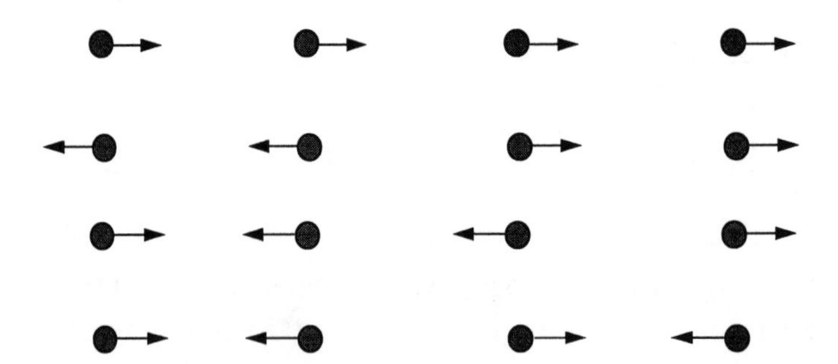

The first represents the low-frequency concerted motion. The second and fourth are symmetric stretches, and the third asymmetric. The frequencies of the different normal modes can be characterized in terms of the number of springs that vibrate. From the diagram we can see that of the five springs, 2, 3, 4, and 5 are vibrating for the first, second, third, and fourth modes, indicating that the normal modes are arranged in order of increasing frequency.

17. The bending modes of the system would look qualitatively like

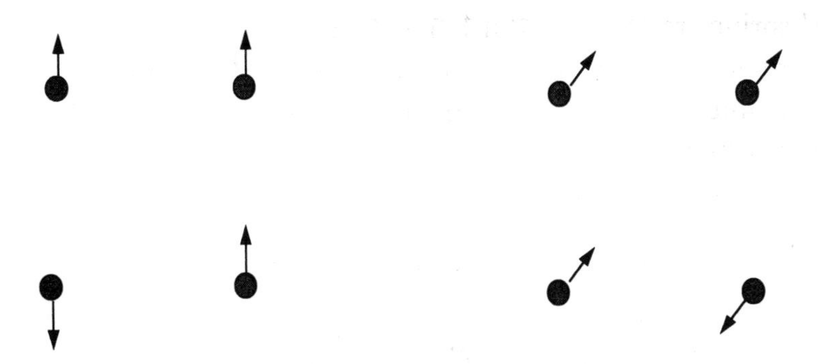

The vertical arrows represent movement in the y-coordinate, and the angled arrows in the z-coordinate. In both coordinates, there is a symmetric mode, in which both particles move together, and an asymmetric mode, in which they move in opposite directions. However, since the y- and z-coordinates are equivalent by symmetry, the two symmetric bends have the same frequency, as do the two asymmetric bends.

18. To simplify the expressions, we use $\sqrt{km} = m\omega$, so that $\beta^2 = m\omega/\hbar$.

$$\Psi_0 = \left(\frac{m\omega}{\hbar\pi}\right)^{1/4} e^{-\frac{m\omega}{2\hbar}x^2}.$$

$$\Psi_1 = \left(\frac{m\omega}{4\hbar\pi}\right)^{1/4} (2\sqrt{\frac{m\omega}{\hbar}}\,x)e^{-\frac{m\omega}{2\hbar}x^2}.$$

$$\Psi_2 = \left(\frac{m\omega}{64\hbar\pi}\right)^{1/4} \left(4\frac{m\omega}{\hbar}x^2 - 2\right) e^{-\frac{m\omega}{2\hbar}x^2}.$$

$$\Psi_3 = \left(\frac{m\omega}{2304\hbar\pi}\right)^{1/4} \left(8\left(\frac{m\omega}{\hbar}\right)^{3/2} x^3 - 12\sqrt{\frac{m\omega}{\hbar}}\,x\right) e^{-\frac{m\omega}{2\hbar}x^2}.$$

$$\Psi_4 = \left(\frac{m\omega}{147456\hbar\pi}\right)^{1/4}\left(16\left(\frac{m\omega}{\hbar}\right)^2 x^4 - 48\frac{m\omega}{\hbar}x^2 + 12\right)e^{-\frac{m\omega}{2\hbar}x^2}.$$

19. $$\frac{d}{dx}\Psi_2 = \left(\frac{m\omega}{64\hbar\pi}\right)^{1/4}\left\{\left[\frac{d}{dx}\left(4\left(\frac{m\omega}{\hbar}\right)^2 x^2 + 2\right)\right]e^{-\frac{m\omega}{2\hbar}x^2} + \left[4\left(\frac{m\omega}{\hbar}\right)^2 x^2 + 2\right]\left[\frac{d}{dx}e^{-\frac{m\omega}{2\hbar}x^2}\right]\right\}$$

$$= \left(\frac{m\omega}{64\hbar\pi}\right)^{1/4}\left(-4\left(\frac{m\omega}{\hbar}\right)^2 x^3 + 10\frac{m\omega}{\hbar}x\right)e^{-\frac{m\omega}{2\hbar}x^2}$$

$$\frac{d^2}{dx^2}\Psi_2 = \left(\frac{m\omega}{64\hbar\pi}\right)^{1/4}\left(4\left(\frac{m\omega}{\hbar}\right)^3 x^4 - 22\left(\frac{m\omega}{\hbar}\right)^2 x^2 + 10\frac{m\omega}{\hbar}\right)e^{-\frac{m\omega}{2\hbar}x^2}$$

$$-\frac{\hbar^2}{2m}\frac{d^2}{dx^2}\Psi_2 = \left(\frac{m\omega}{64\hbar\pi}\right)^{1/4}\left(-2\left(\frac{m^2\omega^3}{\hbar}\right)x^4 + 11m\omega^2x^2 - 5\hbar\omega\right)e^{-\frac{m\omega}{2\hbar}x^2}$$

$$\frac{1}{2}kx^2\,\Psi_2 = \left(\frac{m\omega}{64\hbar\pi}\right)^{1/4}\left(2\left(\frac{m^2\omega^3}{\hbar}\right)x^4 - m\omega^2x^2\right)e^{-\frac{m\omega}{2\hbar}x^2}$$

[Note: We use the relationship $k = \omega^2 m$ to eliminate k in the last step.]
Now, we add the last two equations together to form the complete Hamiltonian operator.

$$\hat{H}\Psi_2 = \left(\frac{m\omega}{64\hbar\pi}\right)^{1/4}\left(10m\omega^2x^2 - 5\hbar\omega\right)e^{-\frac{m\omega}{2\hbar}x^2}$$

$$= \frac{5}{2}\hbar\omega\left(\frac{m\omega}{64\hbar\pi}\right)^{1/4}\left(4\frac{m\omega}{\hbar}x^2 - 2\right)e^{-\frac{m\omega}{2\hbar}x^2} = \frac{5}{2}\hbar\omega\,\Psi_2.$$

$$\frac{d}{dx}\Psi_3 =$$

$$\left(\frac{m\omega}{2304\hbar\pi}\right)^{1/4}\left[-8\left(\frac{m\omega}{\hbar}\right)^{5/2}x^5+36\left(\frac{m\omega}{\hbar}\right)^{3/2}x^3-12\left(\frac{m\omega}{\hbar}\right)^{1/2}x\right]e^{-\frac{m\omega}{2\hbar}x^2}.$$

$$\frac{d^2}{dx^2}\Psi_3 =$$

$$\left(\frac{m\omega}{2304\hbar\pi}\right)^{1/4}\left[8\left(\frac{m\omega}{\hbar}\right)^{7/2}x^5-68\left(\frac{m\omega}{\hbar}\right)^{5/2}x^3+84\left(\frac{m\omega}{\hbar}\right)^{3/2}x\right]e^{-\frac{m\omega}{2\hbar}x^2}.$$

$$-\frac{\hbar^2}{2m}\frac{d^2}{dx^2}\Psi_3 =$$

$$\left(\frac{m\omega}{2304\hbar\pi}\right)^{1/4}\left[-4\left(\frac{m^{5/2}\omega^{7/2}}{\hbar^{3/2}}\right)x^5+34\left(\frac{m^{3/2}\omega^{5/2}}{\hbar^{1/2}}\right)x^3-42(\hbar m)^{1/2}\omega^{3/2}x\right]e^{-\frac{m\omega}{2\hbar}x^2}.$$

$$\frac{1}{2}kx^2\Psi_3 = \left(\frac{m\omega}{2304\hbar\pi}\right)^{1/4}\left[4\left(\frac{m^{5/2}\omega^{7/2}}{\hbar^{3/2}}\right)x^5-6\left(\frac{m^{3/2}\omega^{5/2}}{\hbar^{1/2}}\right)x^3\right]e^{-\frac{m\omega}{2\hbar}x^2}.$$

$$\hat{H}\Psi_3 = \left(\frac{m\omega}{2304\hbar\pi}\right)^{1/4}\left(28\left(\frac{m^{3/2}\omega^{5/2}}{\hbar^{1/2}}\right)x^3-42(\hbar m)^{1/2}\omega^{3/2}\right)e^{-\frac{m\omega}{2\hbar}x^2}$$

$$= \frac{7}{2}\hbar\omega\,\Psi_3.$$

20. Recursion relationship: $h_{n+1}(z) = 2\,z\,h_n(z) - \frac{d\,h_n(z)}{dz}$ [Footnote, p. 251]

$$h_5(z) = 32\,z^5 - 160\,z^3 + 120\,z.$$

$$\frac{d\,h_5(z)}{dz} = 160\,z^4 - 480\,z^2 + 120.$$

$$h_6(z) = 2\,z\,(32\,z^5 - 160\,z^3 + 120\,z) - (160\,z^4 - 480\,z^2 + 120)$$

$$= 64\,z^6 - 480\,z^4 + 720\,z^2 - 120.$$

21. Since $e^0 = 1$, $\psi_n(0)$ is just equal to the product of the normalization constant and the hermite polynomial, $h_n(0)$. Since odd-ordered hermite polynomials have no constant term, $\psi_1(0) = \psi_3(0) = \psi_5(0) = 0$.

In the normalization constant, $\dfrac{N}{\sqrt{2^n n!}}$, N depends on the properties of the oscillator (k and m) but is independent of the vibrational state, n.

$$\Psi_0(0) = \frac{N\,h_0(0)}{\sqrt{2^0\ 0!}} = \frac{N}{1} = N.$$

$$\Psi_2(0) = \frac{N\,h_2(0)}{\sqrt{2^2\ 2!}} = \frac{-2\,N}{\sqrt{8}} = -0.707\ N.$$

$$\Psi_4(0) = \frac{N\,h_4(0)}{\sqrt{2^4\ 4!}} = \frac{12\,N}{\sqrt{384}} = 0.612\ N.$$

$$\Psi_6(0) = \frac{N\,h_6(0)}{\sqrt{2^6\ 6!}} = \frac{-120\,N}{\sqrt{46080}} = -0.559\ N.$$

Chapter 8

Molecular Quantum Mechanics

Exercises

1. The probability density at a particular x value is given by the square of the wavefunction at that point. The values of the n = 0, 2, 4, and 6 harmonic oscillator wavefunctions were found in Problem 7–2. Now, let $N^2 = \beta/(\pi)^{1/2}$, and then

$$P_0(0) = \Psi_0^2(0) = \frac{h_0^2(0)\beta}{2^0 0!\sqrt{\pi}} = \frac{\beta}{\sqrt{\pi}}.$$

$$P_2(0) = \Psi_2^2(0) = \frac{h_2^2(0)\beta}{2^2 2!\sqrt{\pi}} = \frac{\beta}{2\sqrt{\pi}}.$$

$$P_4(0) = \Psi_4^2(0) = \frac{h_4^2(0)\beta}{2^4 4!\sqrt{\pi}} = \frac{3\beta}{8\sqrt{\pi}}.$$

$$P_6(0) = \Psi_6^2(0) = \frac{h_6^2(0)\beta}{2^6 6!\sqrt{\pi}} = \frac{5\beta}{16\sqrt{\pi}}.$$

2. The wavefunction for the n = 1 harmonic oscillator state is

$$\Psi_1 = \left(\frac{m\omega}{4\hbar\pi}\right)^{1/4}\left(2\sqrt{\frac{m\omega}{\hbar}}\,x\right)e^{-\frac{m\omega}{2\hbar}x^2}.$$

The probability density is equal to the square of the wavefunction:

$$P_1 = \left(\frac{m\omega}{4\hbar\pi}\right)^{1/2}\left(\frac{4m\omega}{\hbar}\,x^2\right)e^{-\frac{m\omega}{\hbar}x^2}.$$

At the maximum value of a function, the first derivative must equal 0:

$$\frac{d}{dx}P_1 = \left(\frac{m\omega}{4\hbar\pi}\right)^{1/2}\left(\frac{4m\omega}{\hbar}\right)\left(1-\frac{m\omega}{\hbar}x^2\right)2x\,e^{-\frac{m\omega}{\hbar}x^2} = 0$$

This equation is true if $x = \pm\sqrt{\frac{\hbar}{m\omega}}$, which are the values for which the probability density function has its maximum value.

3. The total energy is given by $E_{tot} = (n_x + 1/2)\,\hbar\omega_x + (n_y + 1/2)\,\hbar\omega_y$. Since $k_x = 4k_y$, then $\omega_x = 2\omega_y$, and the total energy is $E_{tot} = (2n_x + n_y + 3/2)\,\hbar\omega_y$. The following table lists the lowest energy states.

n_x	n_y	E_{tot}
0	0	$3/2\ \hbar\omega$
0	1	$5/2\ \hbar\omega$
0	2	$7/2\ \hbar\omega$
1	0	$7/2\ \hbar\omega$
0	3	$9/2\ \hbar\omega$

n_x	n_y	E_{tot}
1	1	$9/2\ \hbar\omega$
0	4	$11/2\ \hbar\omega$
1	2	$11/2\ \hbar\omega$
2	0	$11/2\ \hbar\omega$

Thus, the degeneracy of the five lowest energy states (with the energy given in parentheses) are 1 ($3/2\ \hbar\omega$), 1 ($5/2\ \hbar\omega$), 2 ($7/2\ \hbar\omega$), 2 ($9/2\ \hbar\omega$), and 3 ($11/2\ \hbar\omega$).

4. Let the mass of the particle be such that $m = h^2/8$, so that $E = n^2 / l^2$ [8-21]. When the length of the box is $l / 2$, then $E = 4n^2 / l^2$, and when the length is $2l$, $E = n^2 / 4 l^2$.

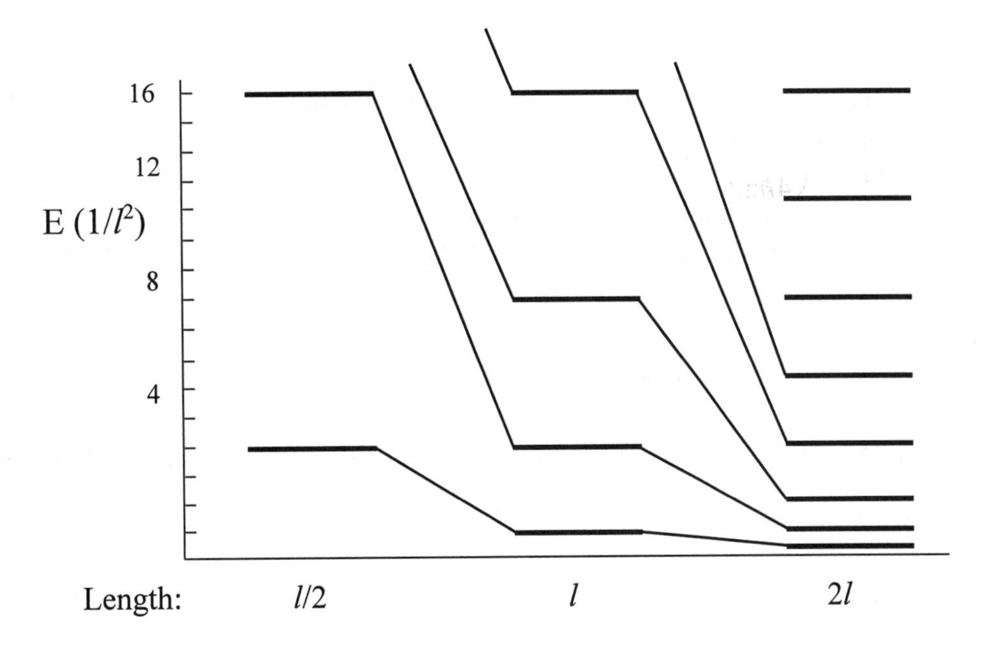

5. We write the total energy in terms of l, where $l_x = 16$, $l_y = 4$, and $l_z = l$:

$$E_{tot} = \frac{n_x^2 h^2}{8m\,l^2} + \frac{n_y^2 h^2}{8m(l/16)^2} + \frac{n_z^2 h^2}{8m(l/4)^2} = \frac{h^2}{8m\,l^2}(n_x^2 + 256n_y^2 + 16n_z^2).$$

$n_x = 1, n_y = 1, n_z = 1, \; E = 273\ h^2/8m\,l^2.$

$n_x = 2, n_y = 1, n_z = 1, \; E = 276\ h^2/8m\,l^2.$

$n_x = 3, n_y = 1, n_z = 1, \; E = 281\ h^2/8m\,l^2.$

$n_x = 4, n_y = 1, n_z = 1, \; E = 288\ h^2/8m\,l^2.$

The degeneracy of each of the first four levels is 1.

6. When coupling more than two sources of angular momentum, the approach is to couple the sources two at a time. In this problem, since $j_1 = 2$ and $j_2 = 1$, the possible values for $j_{12} = j_1 + j_2, j_1 + j_2 - 1, \ldots, |j_1 - j_2| = 3, 2, 1$. To

check this result, we verify that the total number of states is preserved. The total number of uncoupled states = $(2j_1 + 1)(2j_2 + 1) = 5 \times 3 = 15$. The number of coupled states = $(2 \times 3 + 1) + (2 \times 2 + 1) + (2 \times 1 + 1) = 7 + 5 + 3 = 15$.

Now, each of these j_{12} states must be coupled with $j_3 = 1$. When $j_{12} = 3$, the possibilities for $J_{tot} = 3 + 1, \ldots, |3 - 1| = 4, 3, 2$. When $j_{12} = 2$, $J_{tot} = 2 + 1, \ldots, |2 - 1| = 3, 2, 1$. Finally, when $j_{12} = 1$, $J_{tot} = 1 + 1, \ldots, |1 - 1| = 2, 1, 0$. All these possibilities must be included, so that $J_{tot} = 4, 3, 3, 2, 2, 2, 1, 1, 0$. Again, this can be checked by comparing the total number of uncoupled to coupled states (which much match).

Uncoupled: $(2 \times 2 + 1) \times (2 \times 1 + 1) \times (2 \times 1 + 1) = 5 \times 3 \times 3 = 45$.

Coupled: $(2 \times 4 + 1) + 2 \times (2 \times 3 + 1) + 3 \times (2 \times 2 + 1) + 2 \times (2 \times 1 + 1)$
$+ (2 \times 0 + 1) = 9 + 14 + 15 + 6 + 1 = 45$.

Additional Exercises

7. a. $[\hat{B}, \hat{A}^2] = \hat{B}\hat{A}^2 - \hat{A}^2\hat{B} = \frac{d}{dx}x^2 - x^2\frac{d}{dx}$.

To simplify the operator $(d/dx)x^2$, we insert an arbitrary function, f, on which the operator can act, as illustrated on p. 268 in the text.

$$\frac{d}{dx}x^2 f = 2xf + x^2\frac{df}{dx} = \left(2x + \frac{d}{dx}\right)f, \text{ thus } \frac{d}{dx}x^2 = 2x + \frac{d}{dx}$$

$$[\hat{B}, \hat{A}^2] = \frac{d}{dx}x^2 - x^2\frac{d}{dx} = 2x + x^2\frac{d}{dx} - x^2\frac{d}{dx} = 2x.$$

b. $[\hat{B}^2, \hat{A}] = \hat{B}^2\hat{A} - \hat{A}\hat{B}^2 = \frac{d^2}{dx^2}x - x\frac{d^2}{dx^2} = 2\frac{d}{dx} + x\frac{d^2}{dx^2} - x\frac{d^2}{dx^2} = 2\frac{d}{dx}$.

c. $[\hat{B}^2,\hat{A}^2]= \hat{B}^2\hat{A}^2 - \hat{A}^2\hat{B}^2 = \frac{d^2}{dx^2}x^2 - x^2\frac{d^2}{dx^2}$

$$[\hat{B}^2,\hat{A}^2]= 2+4x\frac{d}{dx}+x^2\frac{d^2}{dx^2}-x^2\frac{d^2}{dx^2}= 2+4x\frac{d}{dx}.$$

d. $[\hat{A}^2,\hat{B}^2]= \hat{A}^2\hat{B}^2 - \hat{B}^2\hat{A}^2 = x^2\frac{d^2}{dx^2} - \frac{d^2}{dx^2}x^2$

$$[\hat{A}^2,\hat{B}^2]= x^2\frac{d^2}{dx^2}-\left(2+4x\frac{d}{dx}+x^2\frac{d^2}{dx^2}\right)= -2-4x\frac{d}{dx}.$$

8. $\left[e^{-3x},\frac{d^2}{dx^2}\right]= e^{-3x}\frac{d^2}{dx^2} - \frac{d^2}{dx^2}e^{-3x}$

$$= e^{-3x}\frac{d^2}{dx^2} - \left(9e^{-3x}+2(-3)e^{-3x}\frac{d}{dx}+e^{-3x}\frac{d^2}{dx^2}\right)$$

$$= -9e^{-3x}+6e^{-3x}\frac{d}{dx} = -3e^{-3x}(3-2\frac{d}{dx}).$$

$$\left[e^{-x},\frac{d^2}{dx^2}\right]= e^{-x}\frac{d^2}{dx^2} - \frac{d^2}{dx^2}e^{-x}$$

$$= e^{-x}\frac{d^2}{dx^2} - \left(e^{-x}+2(-1)e^{-x}\frac{d}{dx}+e^{-x}\frac{d^2}{dx^2}\right)$$

$$= -e^{-x}+2e^{-x}\frac{d}{dx} = -e^{-3x}(1-2\frac{d}{dx}).$$

9. We recall the definition of orthogonality, $\langle\Psi_i|\Psi_j\rangle = 0$. Because any constants will enter just as multiplicative factors, they can be neglected.

$$\Psi_0 \propto e^{-z^2},\ \ \Psi_1 \propto ze^{-z^2},\ \text{ and }\ \Psi_2 \propto (4z^2-2)e^{-z^2}$$

where $z = \beta x$ [7-30].

$$\langle\Psi_0|\Psi_1\rangle \propto \int_{-\infty}^{\infty} e^{-z^2/2} z\, e^{-z^2/2}\, dz = 0.$$

This integral is zero because z is an odd function, and the integral of an odd function [a function where f(–x) = –f (x)] from –∞ to ∞ is identically zero.

$$\langle\Psi_1|\Psi_2\rangle \propto \int_{-\infty}^{\infty} z\, e^{-z^2/2}(4z^2-2)\, e^{-z^2/2}\, dz = \int_{-\infty}^{\infty} e^{-z^2/2}(4z^3-2z)\, e^{-z^2/2}\, dz = 0.$$

Likewise, here the polynomial terms are all odd, so the integral is zero.

$$\langle\Psi_0|\Psi_2\rangle \propto \int_{-\infty}^{\infty} e^{-z^2/2}\,(4z^2-2)\, e^{-z^2/2}\, dz = 4\int_{-\infty}^{\infty} z^2 e^{-z^2}\, dz \;-\; 2\int_{-\infty}^{\infty} e^{-z^2}\, dz$$

$$= (4)\frac{1}{2}\sqrt{\pi} - (2)\sqrt{\pi} = 0.$$

[These integrals were evaluated using the formulas in Appendix IV.3, but since those formula are for integrating from 0 to ∞, they are multiplied by 2 for these cases, where the integration is from –∞ to ∞.]

10. Theorem 8.3 [p. 271] states that for a set of functions to be eigenfunctions of two operators, the operators must commute. Since the wavefunctions of the harmonic oscillator are eigenfunctions of the Hamiltonian, the position and momentum operators must commute with the Hamiltonian for the wavefunctions to be eigenfunctions of these operators as well.

$$\left[\hat{H},\hat{x}\right] = \left[-\frac{\hbar^2}{2m}\frac{d^2}{dx^2}, x\right] + \left[\frac{1}{2}kx^2, x\right].$$

$$\left[\hat{H},\hat{x}\right] = \left[-\frac{\hbar^2}{2m}\frac{d^2}{dx^2}x - x\left(-\frac{\hbar^2}{2m}\frac{d^2}{dx^2}\right)\right] + \left((1/2)kx^2 x - x(1/2)kx^2\right)$$

$$= -\frac{\hbar^2}{2m}\left(2\frac{d}{dx} + x\frac{d^2}{dx^2} - x\frac{d^2}{dx^2}\right) + \left[(1/2)kx^3 - (1/2)kx^3\right]$$

$$= -\frac{\hbar^2}{m}\frac{d}{dx} \neq 0.$$

Since the position operator and the Hamiltonian operators do not commute, the wavefunction cannot be an eigenfunction of both.

$$\left[\hat{H}, \hat{p}\right] = \left[-\frac{\hbar^2}{2m}\frac{d^2}{dx^2}, i\hbar\frac{d}{dx}\right] + \left[(1/2)kx^2, i\hbar\frac{d}{dx}\right]$$

$$= -\frac{i\hbar^3}{2m}\left[\left(\frac{d^2}{dx^2}\right)\frac{d}{dx} - \frac{d}{dx}\left(\frac{d^2}{dx^2}\right)\right] + \frac{i\hbar k}{2}\left[x^2\frac{d}{dx} - \frac{d}{dx}x^2\right]$$

$$= -\frac{i\hbar^3}{2m}\left(\frac{d^3}{dx^3} - \frac{d^3}{dx^3}\right) + \frac{i\hbar k}{2}\left(x^2\frac{d}{dx} - \left(2x + x^2\frac{d}{dx}\right)\right)$$

$$= i\hbar kx \neq 0.$$

Like the position operator, the momentum operator does not commute with the Hamiltonian, so the wavefunction cannot be an eigenfunction of both the Hamiltonian and the position operator.

11. $\left\langle \hat{p}^2 \right\rangle_0 = \int_{-\infty}^{\infty} \Psi_0^*(x)\, \hat{p}^2\, \Psi_0(x)\, dx.$

To evaluate this integral, we must first apply the square of the momentum operator to Ψ_0. On p. 252 of the text, the Hamiltonian operator was applied to Ψ_0. We can use that result here, first neglecting the potential energy term ($1/2\ kx^2$) and multiplying by 2m:

$$\hat{p}^2\, \Psi_0 = \hbar^2\beta^2\Psi_0 - \hbar^2\beta^4x^2\Psi_0.$$

Now we can substitute this into the expression above to write

$$\left\langle \hat{p}^2 \right\rangle_0 = \int_{-\infty}^{\infty} \Psi_0^*(x) \left(\hbar^2\beta^2 - \hbar^2\beta^4 x^2\right) \Psi_0(x)\, dx$$

$$= \hbar^2\beta^2 \int_{-\infty}^{\infty} \Psi_0^*(x)\, \Psi_0(x)\, dx - \hbar^2\beta^4 \int_{-\infty}^{\infty} \Psi_0^*(x)\, x^2 \Psi_0(x)\, dx.$$

The first integral is simply equal to $\hbar^2\beta^2$, since the integral is just over the square of the wavefunction, which equals 1 (because the wavefunction is normalized).

The second integral has to be evaluated by substituting in Ψ_0:

$$-\hbar^2\beta^4 \int_{-\infty}^{\infty} \left(\frac{\beta^2}{\pi}\right)^{1/4} e^{-\beta^2 x^2/2} x^2 \left(\frac{\beta^2}{\pi}\right)^{1/4} e^{-\beta^2 x^2/2}\, dx$$

$$= -\frac{\hbar^2\beta^5}{\sqrt{\pi}} \int_{-\infty}^{\infty} x^2\, e^{-\beta^2 x^2}\, dx = -\frac{\hbar^2\beta^5}{\sqrt{\pi}} \left(\frac{1}{2}\sqrt{\frac{\pi}{\beta^6}}\right) = -\frac{\hbar^2\beta^2}{2}.$$

$$\left\langle \hat{p}^2 \right\rangle_0 = \hbar^2\beta^2 - \frac{\hbar^2\beta^2}{2} = \frac{\hbar^2\beta^2}{2}.$$

For the n = 1 wavefunction,

$$\left\langle \hat{p}^2 \right\rangle_1 = \int_{-\infty}^{\infty} \Psi_1^*(x)\, \hat{p}^2\, \Psi_1(x)\, dx.$$

$$\Psi_1 = \left(\frac{\beta^2}{4\pi}\right)^{1/4} (2\beta\, x) e^{-\beta^2 x^2/2}.$$

$$\hat{p}^2 \Psi_1 = -\hbar^2 \frac{d^2}{dx^2} \left(\frac{\beta^2}{4\pi}\right)^{1/4} (2\beta\, x) e^{-\beta^2 x^2/2}$$

$$= -2\hbar^2\beta \left(\frac{\beta^2}{4\pi}\right)^{1/4} \left(-3\beta^2 x + \beta^4 x^3\right) e^{-\beta^2 x^2/2}.$$

$$\langle \hat{p}^2 \rangle_1 = -2\hbar^2\beta^3\left(\frac{\beta^2}{4\pi}\right)^{1/2}\int_{-\infty}^{\infty}(2\beta\, x)e^{-\beta^2x^2/2}\left(-3x+\beta^2x^3\right)e^{-\beta^2x^2/2}\,dx$$

$$= -\frac{2\hbar^2\beta^5}{\sqrt{\pi}}\left[-3\int_{-\infty}^{\infty}x^2e^{-\beta^2x^2}\,dx + \beta^2\int_{-\infty}^{\infty}x^4e^{-\beta^2x^2}\,dx\right]$$

$$= -\frac{2\hbar^2\beta^5}{\sqrt{\pi}}\left[-3\left(\frac{1}{2}\sqrt{\frac{\pi}{\beta^6}}\right)+\beta^2\left(\frac{3}{4}\sqrt{\frac{\pi}{\beta^{10}}}\right)\right]$$

$$= 3\hbar^2\beta^2 - \frac{3}{2}\hbar^2\beta^2 = \frac{3}{2}\hbar^2\beta^2.$$

12. $\Delta p_0 = \sqrt{\langle \hat{p}^2 \rangle_0 - \langle \hat{p} \rangle_0^2}$ [8-11].

The average value of the momentum operator squared was determined in the previous problem. The average value of the momentum operator itself is identically zero, because the derivative operation is effectively an odd function [convince yourself that x^2 (d/dx) x^2 is an odd function]. Therefore,

$$\Delta p_0 = \sqrt{\langle \hat{p}^2 \rangle_0 - \langle \hat{p} \rangle_0^2} = \sqrt{\langle \hat{p}^2 \rangle_0} = \sqrt{\frac{\hbar^2\beta^2}{2}} = \frac{\hbar\beta}{\sqrt{2}}.$$

$$\Delta p_1 = \sqrt{\langle \hat{p}^2 \rangle_1 - \langle \hat{p} \rangle_1^2} = \sqrt{\langle \hat{p}^2 \rangle_1} = \sqrt{\frac{3}{2}\hbar^2\beta^2} = \sqrt{\frac{3}{2}}\hbar\beta.$$

13. $\langle x^2 \rangle_0 = \int_{-\infty}^{\infty}\Psi_0(x)\, x^2\, \Psi_0(x)\, dx$

$$= \int_{-\infty}^{\infty}\left(\frac{\beta^2}{\pi}\right)^{1/4} e^{-\beta^2x^2/2}x^2\left(\frac{\beta^2}{\pi}\right)^{1/4} e^{-\beta^2x^2/2}\,dx$$

$$= \frac{\beta}{\sqrt{\pi}}\int_{-\infty}^{\infty}x^2\, e^{-\beta^2x^2}dx = \frac{\beta}{\sqrt{\pi}}\left(\frac{1}{2}\sqrt{\frac{\pi}{\beta^6}}\right) = \frac{1}{2\beta^2}.$$

$$\langle x^2\rangle_1 = \int_{-\infty}^{\infty}\Psi_1(x)\, x^2\, \Psi_1(x)\, dx$$

$$= \int_{-\infty}^{\infty}\left(\frac{\beta^2}{4\pi}\right)^{1/4}(2\beta\, x)e^{-\beta^2x^2/2}x^2\left(\frac{\beta^2}{4\pi}\right)^{1/4}(2\beta\, x)e^{-\beta^2x^2/2}\, dx$$

$$= \frac{2\beta^3}{\sqrt{\pi}}\int_{-\infty}^{\infty}x^4e^{-\beta^2x^2}\, dx = \frac{2\beta^3}{\sqrt{\pi}}\left(\frac{3}{4}\sqrt{\frac{\pi}{\beta^{10}}}\right) = \frac{3}{2\beta^2}.$$

$$\langle x^2\rangle_2 = \int_{-\infty}^{\infty}\Psi_2(x)\, x^2\, \Psi_2(x)\, dx$$

$$= \int_{-\infty}^{\infty}\left(\frac{\beta^2}{64\pi}\right)^{1/4}(4\beta^2x^2 - 2)e^{-\beta^2x^2/2}x^2\left(\frac{\beta^2}{64\pi}\right)^{1/4}(4\beta^2x^2 - 2)e^{-\beta^2x^2/2}\, dx$$

$$= \frac{\beta}{8\sqrt{\pi}}\int_{-\infty}^{\infty}\left(16\beta^4x^6 - 16\beta^2x^4 + 4x^2\right)e^{-\beta^2x^2}\, dx$$

$$= \frac{\beta}{8\sqrt{\pi}}\left(16\beta^4\,\frac{15}{8}\sqrt{\frac{\pi}{\beta^{14}}} - 16\beta^2\,\frac{3}{4}\sqrt{\frac{\pi}{\beta^{10}}} + 4\,\frac{1}{2}\sqrt{\frac{\pi}{\beta^6}}\right) = \frac{5}{2\beta^2}.$$

$$\langle x^2\rangle_3 = \int_{-\infty}^{\infty}\Psi_3(x)\, x^2\, \Psi_3(x)\, dx$$

$$= \int_{-\infty}^{\infty}\left(\frac{\beta^2}{48^2\,\pi}\right)^{1/4}(8\beta^3x^3 - 12\,\beta x)e^{-\beta^2x^2/2}x^2$$

$$\times\left(\frac{\beta^2}{48^2\,\pi}\right)^{1/4}(8\beta^3x^3 - 12\,\beta x)e^{-\beta^2x^2/2}\, dx$$

$$= \frac{\beta}{48\sqrt{\pi}}\int_{-\infty}^{\infty}\left(64\beta^6x^8 - 192\beta^4x^6 + 144\beta^2x^4\right)e^{-\beta^2x^2}\, dx$$

$$= \frac{\beta}{48\sqrt{\pi}}\left(64\beta^6 \frac{105}{16}\sqrt{\frac{\pi}{\beta^{18}}} - 192\beta^4 \frac{15}{8}\sqrt{\frac{\pi}{\beta^{14}}} + 144\beta^2 \frac{3}{4}\sqrt{\frac{\pi}{\beta^{10}}}\right) = \frac{7}{2\beta^2}.$$

$\langle x^2 \rangle_n = (2n + 1)/2\beta^2$.

14. $\Delta x = \sqrt{\langle \hat{x}^2 \rangle - \langle \hat{x} \rangle^2} = \sqrt{\frac{2n+1}{2\beta^2} - 0} = \frac{\sqrt{n+(1/2)}}{\beta}.$

15. The total energy is given by

$$E_{tot} = (n_x + 1/2)\,\hbar\omega + (n_y + 1/2)\,\hbar\omega + (n_z + 1/2)\,\hbar\omega$$

$$= 3/2\,\hbar\omega + (n_x + n_y + n_z)\,\hbar\omega.$$

E_{tot}	Degeneracy	n_x	n_y	n_z
$3/2\,\hbar\omega$	1	0	0	0
$5/2\,\hbar\omega$	3	1	0	0
		0	1	0
		0	0	1
$7/2\,\hbar\omega$	6	2	0	0
		1	1	0
		1	0	0
		0	2	0
		0	1	1
		0	0	2
$9/2\,\hbar\omega$	10	3	0	0
		2	1	0
		2	0	1
		1	2	0
		1	1	1
		1	0	2
		0	3	0
		0	2	1
		0	1	2
		0	0	3

16. Because both the n and m quantum numbers enter the energy expression as squares, the sign doesn't matter. The following is a complete list of the energies of all possible states. The degeneracy of each energy level is given with the first entry of that energy.

n	m	E ($\hbar$a)	Degeneracy
1	±2	0	4
1	±1	−3/4	2
1	0	−1	1
2	±3	2	2
2	±2	3/4	2
2	±1	0	
2	0	−1/4	1
3	±4	35/9	2
3	±3	77/36	2
3	±2	8/9	2
3	±1	5/36	2
3	0	−1/9	1

17. The one-dimensional particle in a box wavefunction is [8-24]

$$\Psi_n(x) = \sqrt{\frac{2}{l}}\sin\left(\frac{n\pi x}{l}\right).$$

In general, the expectation value of the position, x, is

$$\langle x\rangle_n = \int_0^l \sqrt{\frac{2}{l}}\sin\left(\frac{n\pi x}{l}\right) x \sqrt{\frac{2}{l}}\sin\left(\frac{n\pi x}{l}\right) dx = \frac{2}{l}\int_0^l x\,\sin^2\left(\frac{n\pi x}{l}\right) dx$$

$$= \frac{2}{l}\left[\frac{x^2}{4} - \frac{x\,l\sin\left(\frac{2n\pi x}{l}\right)}{(4n\pi)} - \frac{l^2\cos\left(\frac{2n\pi x}{l}\right)}{8(n\pi)^2}\right]_0^l = \left(\frac{2}{l}\right)\left(\frac{l^2}{4}\right) = \frac{l}{2}.$$

Note that $\langle x\rangle_n$ is independent of n. Determination of Δx also requires $\langle x^2\rangle_n$:

$$\langle x^2\rangle_n = \int_0^l \sqrt{\frac{2}{l}}\sin(\frac{n\pi x}{l})\ x^2\ \sqrt{\frac{2}{l}}\sin(\frac{n\pi x}{l})\ dx = \frac{2}{l}\int_0^l x^2\ \sin^2(\frac{n\pi x}{l})\ dx$$

$$= \frac{2}{l}\left[\frac{x^3}{6} - \left(\frac{x^2 l}{(4n\pi)} - \frac{l^3}{8(n\pi)^3}\right)\sin(\frac{2n\pi x}{l}) - \frac{x\,l^2\cos(\frac{2n\pi x}{l})}{4(n\pi)^2}\right]_0^l = \frac{l^2}{3}.$$

Again, $\langle x^2\rangle_n$ is independent of n. So for all n,

$$\Delta x = \sqrt{\langle x^2\rangle - \langle x\rangle^2} = \sqrt{\frac{l^2}{3} - \frac{l^2}{4}} = \frac{l}{\sqrt{12}}.$$

18. Using the relationship that $(1/2i)\ (e^{i\theta} - e^{-i\theta}) = \sin\theta$, we have

$$\Psi_n(x) = \sqrt{\frac{2}{l}}\sin\left(\frac{n\pi x}{l}\right) = \sqrt{\frac{2}{l}}\left(\frac{1}{2i}\right)\left(e^{in\pi x/l} - e^{-in\pi x/l}\right).$$

$$\hat{p} = i\hbar\frac{d}{dx}.$$

$$\hat{p}\,\Psi_n = i\hbar\frac{1}{i\sqrt{2l}}\ \frac{d}{dx}\ \left(e^{in\pi x/l} - e^{-in\pi x/l}\right)$$

$$= \frac{\hbar}{\sqrt{2l}}\left(\frac{in\pi}{l}e^{in\pi x/l} + \frac{in\pi}{l}e^{-in\pi x/l}\right)$$

$$= \frac{i\hbar n\pi}{\sqrt{2}l^{3/2}}\left(e^{in\pi x/l} + e^{-in\pi x/l}\right).$$

$$\langle\hat{p}\rangle_n = \int_0^l \Psi_n^*(x)\,\hat{p}\,\Psi_n(x)\,dx$$

$$= \int_0^l\left(-\frac{1}{i\sqrt{2l}}\right)\left(e^{-in\pi x/l} - e^{in\pi x/l}\right)\frac{i\hbar n\pi}{\sqrt{2}l^{3/2}}\left(e^{in\pi x/l} + e^{-in\pi x/l}\right)dx.$$

$$= -\frac{\hbar n\pi}{2l^2}\int_0^l\left(e^{-2in\pi x/l} - e^{2in\pi x/l}\right)dx = \frac{\hbar}{4i}\left(e^{-2in\pi x/l} + e^{2in\pi x/l}\right)_0^l = 0.$$

$\langle p \rangle_n = 0$ for all values of n.

To calculate Δp, we first need $\langle p^2 \rangle_n$.

$$\hat{p}^2 \Psi_n = -\hbar^2 \frac{1}{i\sqrt{2l}} \frac{d^2}{dx^2}\left(e^{in\pi x/l} - e^{-in\pi x/l}\right)$$

$$= \frac{\hbar^2}{i\sqrt{2l}}\left(\left[\frac{in\pi}{l}\right]^2 e^{in\pi x/l} - \left[\frac{in\pi}{l}\right]^2 e^{-in\pi x/l}\right)$$

$$= -\frac{\hbar^2 n^2 \pi^2}{i\sqrt{2}l^{5/2}}\left(e^{in\pi x/l} - e^{-in\pi x/l}\right).$$

$$\left\langle \hat{p}^2 \right\rangle_n = \int_0^l \Psi_n^*(x)\, \hat{p}\, \Psi_n(x)\, dx$$

$$= \int_0^l -\frac{1}{i\sqrt{2l}}\left(e^{-in\pi x/l} - e^{in\pi x/l}\right)\left(-\frac{\hbar^2 n^2 \pi^2}{i\sqrt{2}l^{5/2}}\right)\left(e^{in\pi x/l} - e^{-in\pi x/l}\right) dx.$$

$$= \left(-\frac{\hbar^2 n^2 \pi^2}{2\, l^3}\right)\int_0^l \left(2 - \left[e^{2in\pi x/l} + e^{-2in\pi x/l}\right]\right) dx = \left(-\frac{\hbar^2 n^2 \pi^2}{l^2}\right).$$

$$\Delta p = \sqrt{\left\langle p^2 \right\rangle - \langle p \rangle^2} = \sqrt{\left\langle p^2 \right\rangle} = \sqrt{-\frac{\hbar^2 n^2 \pi^2}{l^2}} = \frac{i\hbar n\pi}{l}.$$

$\Delta p_0 = 0.$

$\Delta p_1 = i\hbar p / 1.$

$\Delta p_2 = i\hbar 2p / 1.$

19. The wavefunctions are shown below with their baselines incremented for clarity. Each wavefunction is identically zero at the point where the potential becomes infinite. The functions then behave similarly to the harmonic oscillator wavefunctions, oscillating at higher frequency at higher states before decaying to zero as the potential energy becomes greater than the vibrational state energy.

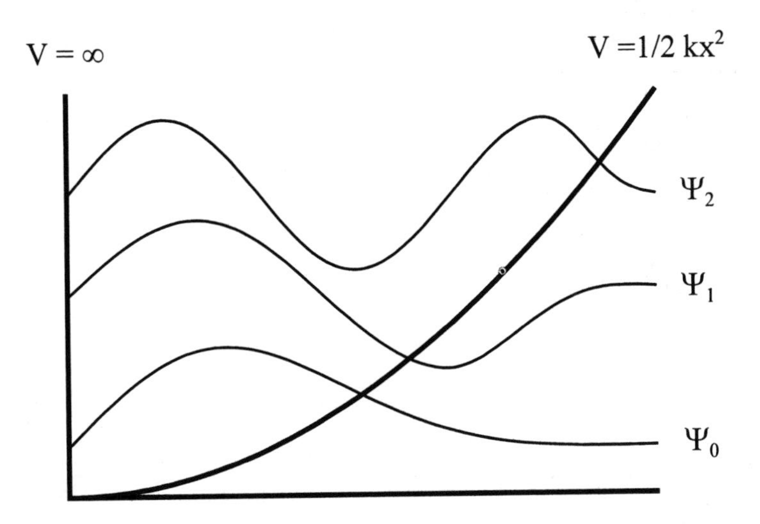

20. $E_n(J) = n^2h^2 / 8ml^2$. To convert to units of cm^{-1}, we use the relationship on p. 634:

$$E_n\ (cm^{-1}) = E_n\ (J) / hc = n^2h / 8mcl^2.$$

In these units, the energy difference between the n = 1 (ground) and n = 2 (first excited state) is given by

$$\Delta E_{12}(cm^{-1}) = (2^2 - 1^2)\ h / 8mcl^2 = 3h / 8mcl^2.$$

Setting $\Delta E_{12} = 10{,}000\ cm^{-1}$, we can solve for l.

$$l = \sqrt{\frac{(3)(6.626 \times 10^{-34}\ \text{J s})}{(8)(9.109 \times 10^{-31}\ \text{kg})\left(2.998 \times 10^{10}\ \text{cm s}^{-1}\right)(10{,}000\text{cm}^{-1})}} = 9.54 \times 10^{-10}\ \text{m}$$
$$= 9.54\ \text{Å}.$$

Since the bond length of a carbon–carbon triple bond is roughly 1.2 Å and that of the carbon–carbon single bond is about 1.5 Å, a linear polyene of a 3 subunits, $H\text{–}(C{\equiv}C\text{–}C)_3{\equiv}C\text{–}H$, would be required to match this length.

21. The energy for the three-dimensional particle in a cubic box is

$$E_{tot} = (n_x^{\ 2} + n_y^{\ 2} + n_z^{\ 2})\ h^2/8ml^2.$$

Energy	Degeneracy	States (n_x,n_y,n_z)
$3h^2/8ml^2$	1	(1,1,1)
$6h^2/8ml^2$	3	(2,1,1), (1,2,1), (1,1,2)
$9h^2/8ml^2$	3	(2,2,1), (2,1,2), (1,2,2)
$11h^2/8ml^2$	3	(3,1,1), (1,3,1), (1,1,3)
$12h^2/8ml^2$	1	(2,2,2)

22. [8-48]:

$$[\hat{L}_x,\hat{L}_y] = -\hbar^2\left[\left(y\frac{\partial}{\partial z}-z\frac{\partial}{\partial y}\right)\left(z\frac{\partial}{\partial x}-x\frac{\partial}{\partial z}\right)-\left(z\frac{\partial}{\partial x}-x\frac{\partial}{\partial z}\right)\left(y\frac{\partial}{\partial z}-z\frac{\partial}{\partial y}\right)\right]$$

$$= -\hbar^2\left[y\frac{\partial}{\partial x}+yz\frac{\partial}{\partial z}\frac{\partial}{\partial x}-yx\frac{\partial^2}{\partial z^2}-z^2\frac{\partial}{\partial y}\frac{\partial}{\partial x}+zx\frac{\partial}{\partial y}\frac{\partial}{\partial z}\right.$$

$$\left.-\left(zy\frac{\partial}{\partial x}\frac{\partial}{\partial z}-z^2\frac{\partial}{\partial x}\frac{\partial}{\partial y}-xy\frac{\partial^2}{\partial z^2}+x\frac{\partial}{\partial y}+xz\frac{\partial}{\partial z}\frac{\partial}{\partial y}\right)\right]$$

$$= -\hbar^2\left(y\frac{\partial}{\partial x}-x\frac{\partial}{\partial y}\right) = (i\hbar)(-i\hbar)\left(x\frac{\partial}{\partial y}-y\frac{\partial}{\partial x}\right) = i\hbar\hat{L}_z.$$

[8-49]:

$$[\hat{L}_y,\hat{L}_z] = -\hbar^2\left[\left(z\frac{\partial}{\partial x}-x\frac{\partial}{\partial z}\right)\left(x\frac{\partial}{\partial y}-y\frac{\partial}{\partial x}\right)-\left(x\frac{\partial}{\partial y}-y\frac{\partial}{\partial x}\right)\left(z\frac{\partial}{\partial x}-x\frac{\partial}{\partial z}\right)\right]$$

$$= -\hbar^2\left[z\frac{\partial}{\partial y}+zx\frac{\partial}{\partial x}\frac{\partial}{\partial y}-zy\frac{\partial^2}{\partial x^2}-x^2\frac{\partial}{\partial z}\frac{\partial}{\partial y}+xy\frac{\partial}{\partial z}\frac{\partial}{\partial x}\right.$$

$$\left.-\left(xz\frac{\partial}{\partial y}\frac{\partial}{\partial x}-x^2\frac{\partial}{\partial y}\frac{\partial}{\partial z}-yz\frac{\partial^2}{\partial x^2}+y\frac{\partial}{\partial z}+yx\frac{\partial}{\partial x}\frac{\partial}{\partial z}\right)\right]$$

$$= -\hbar^2\left(z\frac{\partial}{\partial y} - y\frac{\partial}{\partial z}\right) = (i\hbar)(-i\hbar)\left(y\frac{\partial}{\partial z} - z\frac{\partial}{\partial y}\right) = i\hbar\hat{L}_x.$$

[8-50]:

$$[\hat{L}_z, \hat{L}_x] = -\hbar^2\left[\left(x\frac{\partial}{\partial y} - y\frac{\partial}{\partial x}\right)\left(y\frac{\partial}{\partial z} - z\frac{\partial}{\partial y}\right) - \left(y\frac{\partial}{\partial z} - z\frac{\partial}{\partial y}\right)\left(x\frac{\partial}{\partial y} - y\frac{\partial}{\partial x}\right)\right]$$

$$= -\hbar^2\left[x\frac{\partial}{\partial z} + xy\frac{\partial}{\partial y}\frac{\partial}{\partial z} - xz\frac{\partial^2}{\partial y^2} - y^2\frac{\partial}{\partial x}\frac{\partial}{\partial z} + yz\frac{\partial}{\partial x}\frac{\partial}{\partial y}\right.$$

$$\left. - \left(yx\frac{\partial}{\partial z}\frac{\partial}{\partial y} - y^2\frac{\partial}{\partial z}\frac{\partial}{\partial x} - zx\frac{\partial^2}{\partial y^2} + z\frac{\partial}{\partial x} + zy\frac{\partial}{\partial y}\frac{\partial}{\partial x}\right)\right]$$

$$= -\hbar^2\left(x\frac{\partial}{\partial z} - z\frac{\partial}{\partial x}\right) = (i\hbar)(-i\hbar)\left(z\frac{\partial}{\partial x} - x\frac{\partial}{\partial z}\right) = i\hbar\hat{L}_y.$$

23. $\hat{L}^2 = -\hbar^2\left(\frac{1}{\sin\theta}\frac{\partial}{\partial\theta}\sin\theta\frac{\partial}{\partial\theta} + \frac{1}{\sin^2\theta}\frac{\partial^2}{\partial\phi^2}\right).$ [8-52]

$\hat{L}_x = -i\hbar\left(-\sin\phi\frac{\partial}{\partial\theta} - \frac{\cos\phi}{\tan\theta}\frac{\partial}{\partial\phi}\right).$ [8-51]

$$\hat{L}^2\,\hat{L}_x = -i\hbar^3\left[\frac{1}{\sin\theta}\frac{\partial}{\partial\theta}\sin\theta\frac{\partial}{\partial\theta}\left(\sin\phi\frac{\partial}{\partial\theta} + \frac{\cos\phi}{\tan\theta}\frac{\partial}{\partial\phi}\right)\right.$$

$$\left. + \frac{1}{\sin^2\theta}\frac{\partial^2}{\partial\phi^2}\left(\sin\phi\frac{\partial}{\partial\theta} + \frac{\cos\phi}{\tan\theta}\frac{\partial}{\partial\phi}\right)\right]$$

$$= \left(\sin\phi\frac{\partial^3}{\partial\theta^3} + \frac{\sin\phi}{\tan\theta}\frac{\partial^2}{\partial\theta^2} - \frac{\cos\phi}{\sin^2\theta}\frac{\partial}{\partial\theta}\frac{\partial}{\partial\phi} + \frac{\cos\phi}{\tan\theta}\frac{\partial^2}{\partial\theta^2}\frac{\partial}{\partial\phi} + \frac{\cos\phi}{\sin^2\theta\tan\theta}\frac{\partial}{\partial\phi}\right.$$

$$- \cos\phi \frac{\partial}{\partial\theta}\frac{\partial}{\partial\phi} - \frac{\sin\phi}{\sin^2\theta}\frac{\partial}{\partial\theta} + \frac{2\cos\phi}{\sin^2\theta}\frac{\partial}{\partial\theta}\frac{\partial}{\partial\phi} + \frac{\sin\phi}{\sin^2\theta}\frac{\partial^2}{\partial\phi^2}\frac{\partial}{\partial\theta}$$

$$\left. - \frac{\cos\phi}{\sin^2\theta\tan\theta}\frac{\partial}{\partial\phi} - \frac{2\sin\phi}{\sin^2\theta\tan\theta}\frac{\partial^2}{\partial\phi^2} + \frac{\cos\phi}{\sin^2\theta\tan\theta}\frac{\partial^3}{\partial\phi^3}\right).$$

$$\hat{L}_x\hat{L}^2 = i\hbar^3\left[\sin\phi\frac{\partial}{\partial\theta}\left(\frac{1}{\sin\theta}\frac{\partial}{\partial\theta}\sin\theta\frac{\partial}{\partial\theta} + \frac{1}{\sin^2\theta}\frac{\partial^2}{\partial\phi^2}\right)\right.$$

$$\left. + \frac{\cos\phi}{\tan\theta}\frac{\partial}{\partial\phi}\left(\frac{1}{\sin\theta}\frac{\partial}{\partial\theta}\sin\theta\frac{\partial}{\partial\theta} + \frac{1}{\sin^2\theta}\frac{\partial^2}{\partial\phi^2}\right)\right]$$

$$= \left(-\frac{\sin\phi}{\sin^2\theta}\frac{\partial}{\partial\theta} + \frac{\sin\phi}{\tan\theta}\frac{\partial^2}{\partial\theta^2} + \sin\phi\frac{\partial^3}{\partial\theta^3} + \frac{\sin\phi}{\sin^2\theta}\frac{\partial}{\partial\theta}\frac{\partial^2}{\partial\phi^2} - \frac{2\sin\phi}{\sin^2\theta\tan\theta}\frac{\partial^2}{\partial\phi^2}\right.$$

$$\left. + \frac{\cos\phi}{\tan^2\theta}\frac{\partial}{\partial\phi}\frac{\partial}{\partial\theta} + \frac{\cos\phi}{\tan\theta}\frac{\partial}{\partial\phi}\frac{\partial^2}{\partial\theta^2} + \frac{\cos\phi}{\sin^2\theta\tan^2\theta}\frac{\partial^3}{\partial\phi^3}\right).$$

Comparing term by term, we see that $\hat{L}^2\hat{L}_x - \hat{L}_x\hat{L}^2 = 0$.

$$\hat{L}_y = -i\hbar\left(\cos\phi\frac{\partial}{\partial\theta} - \frac{\sin\phi}{\tan\theta}\frac{\partial}{\partial\phi}\right). \qquad [8\text{-}51]$$

$$\hat{L}^2\,\hat{L}_y = i\hbar^3\left[\frac{1}{\sin\theta}\frac{\partial}{\partial\theta}\sin\theta\frac{\partial}{\partial\theta}\left(\cos\phi\frac{\partial}{\partial\theta} - \frac{\sin\phi}{\tan\theta}\frac{\partial}{\partial\phi}\right)\right.$$

$$\left. + \frac{1}{\sin^2\theta}\frac{\partial^2}{\partial\phi^2}\left(\cos\phi\frac{\partial}{\partial\theta} - \frac{\sin\phi}{\tan\theta}\frac{\partial}{\partial\phi}\right)\right]$$

$$= \left(\cos\phi\frac{\partial^3}{\partial\theta^3} + \frac{\cos\phi}{\tan\theta}\frac{\partial^2}{\partial\theta^2} + \frac{\sin\phi}{\sin^2\theta}\frac{\partial}{\partial\theta}\frac{\partial}{\partial\phi} - \frac{\sin\phi}{\tan\theta}\frac{\partial^2}{\partial\theta^2}\frac{\partial}{\partial\phi} - \frac{\sin\phi}{\sin^2\theta\tan\theta}\frac{\partial}{\partial\phi}\right.$$

$$+\sin\phi\frac{\partial}{\partial\theta}\frac{\partial}{\partial\phi}-\frac{\cos\phi}{\sin^2\theta}\frac{\partial}{\partial\theta}-\frac{2\sin\phi}{\sin^2\theta}\frac{\partial}{\partial\theta}\frac{\partial}{\partial\phi}+\frac{\cos\phi}{\sin^2\theta}\frac{\partial^2}{\partial\phi^2}\frac{\partial}{\partial\theta}$$

$$+\frac{\sin\phi}{\sin^2\theta\tan\theta}\frac{\partial}{\partial\phi}-\frac{2\cos\phi}{\sin^2\theta\tan\theta}\frac{\partial^2}{\partial\phi^2}-\frac{\sin\phi}{\sin^2\theta\tan\theta}\frac{\partial^3}{\partial\phi^3}\Bigg).$$

$$\hat{L}_y\hat{L}^2 = i\hbar^3\Bigg[\cos\phi\frac{\partial}{\partial\theta}\left(\frac{1}{\sin\theta}\frac{\partial}{\partial\theta}\sin\theta\frac{\partial}{\partial\theta}+\frac{1}{\sin^2\theta}\frac{\partial^2}{\partial\phi^2}\right)$$

$$-\frac{\sin\phi}{\tan\theta}\frac{\partial}{\partial\phi}\left(\frac{1}{\sin\theta}\frac{\partial}{\partial\theta}\sin\theta\frac{\partial}{\partial\theta}+\frac{1}{\sin^2\theta}\frac{\partial^2}{\partial\phi^2}\right)\Bigg]$$

$$=\Bigg(-\frac{\cos\phi}{\sin^2\theta}\frac{\partial}{\partial\theta}+\frac{\cos\phi}{\tan\theta}\frac{\partial^2}{\partial\theta^2}+\cos\phi\frac{\partial^3}{\partial\theta^3}+\frac{\cos\phi}{\sin^2\theta}\frac{\partial}{\partial\theta}\frac{\partial^2}{\partial\phi^2}-\frac{2\cos\phi}{\sin^2\theta\tan\theta}\frac{\partial^2}{\partial\phi^2}$$

$$-\frac{\sin\phi}{\tan^2\theta}\frac{\partial}{\partial\phi}\frac{\partial}{\partial\theta}-\frac{\sin\phi}{\tan\theta}\frac{\partial}{\partial\phi}\frac{\partial^2}{\partial\theta^2}-\frac{\sin\phi}{\sin^2\theta\tan^2\theta}\frac{\partial^3}{\partial\phi^3}\Bigg).$$

Comparing term by term, you can see that $\hat{L}^2\hat{L}_y-\hat{L}_y\hat{L}^2=0$.

24. $\hat{L}^2\sin^m\theta\cos\theta\, e^{im\phi}=$

$$-\hbar^2\left(\frac{1}{\sin\theta}\frac{\partial}{\partial\theta}\sin\theta\frac{\partial}{\partial\theta}+\frac{1}{\sin^2\theta}\frac{\partial^2}{\partial\phi^2}\right)\sin^m\theta\cos\theta\, e^{im\phi}$$

$$=-\hbar^2\Bigg\{\left[\frac{1}{\sin\theta}\frac{\partial}{\partial\theta}\sin\theta\left(m\sin^{m-1}\theta\cos^2\theta-\sin^{m+1}\theta\right)e^{im\phi}\right]$$

$$+\frac{1}{\sin^2\theta}\sin^m\theta\cos\theta\left(-m^2\right)e^{im\phi}\Bigg\}$$

$$=-\hbar^2\left\{\left[\frac{1}{\sin\theta}\frac{\partial}{\partial\theta}\left(m\sin^m\theta\cos^2\theta-\sin^{m+2}\theta\right)\right]-m^2\sin^{m-2}\theta\cos\theta\right\}e^{im\phi}$$

$$= -\hbar^2\Big(m^2 \sin^{m-2}\theta\cos^3\theta - 2m\ \sin^m\theta\ \cos\theta - (m+2)\sin^m\theta\ \cos\theta$$
$$- m^2 \sin^{m-2}\theta\ \cos\theta\Big)e^{im\phi}.$$

Now we can eliminate the $\cos^3$ term by substituting $\cos^2 = 1 - \sin^2$ to get

$$= -\hbar^2(-m^2 - 3m - 2)\sin^m\theta\ \cos\theta\ e^{im\phi}$$

$$= \hbar^2(m+1)(m+2)\sin^m\theta\ \cos\theta\ e^{im\phi}.$$

The eigenvalue is $\hbar^2(m+1)(m+2)$.

25. $Y_{20} = \Theta_{20}\,\Phi_0$ [8-59]

$$= \sqrt{\frac{(2(2)+1)(2-0)!}{2(2+0)!}}\,P_2^0(\cos\theta)\,\Phi_0$$

$$= \frac{1}{4}\sqrt{\frac{5}{\pi}}\,(3\cos^3\theta - 1).$$

$$\hat{L}_x Y_{20} = -i\hbar\left(-\sin\phi\frac{\partial}{\partial\theta} - \frac{\cos\phi}{\tan\theta}\frac{\partial}{\partial\phi}\right)\left(\frac{1}{4}\sqrt{\frac{5}{\pi}}\,(3\cos^2\theta - 1)\right)$$

$$= i\hbar\frac{1}{4}\sqrt{\frac{5}{\pi}}\sin\phi\frac{\partial}{\partial\theta}\ (3\cos^2\theta - 1)$$

$$= -i\hbar\frac{3}{2}\sqrt{\frac{5}{\pi}}\sin\phi\cos\theta\sin\theta.$$

$$\left\langle Y_{20}|\hat{L}_x Y_{20}\right\rangle = \int_0^{2\pi}\int_0^{\pi}\left(\frac{1}{4}\sqrt{\frac{5}{\pi}}\,(3\cos^2\theta - 1)\right)\left(-i\hbar\frac{3}{2}\sqrt{\frac{5}{\pi}}\sin\phi\cos\theta\sin\theta\right)\sin\theta\,d\theta\,d\phi$$

$$= \frac{-i\hbar 15}{8\pi}\left[\int_0^{\pi} 3\cos^3\theta\sin^2\theta\,d\theta\int_0^{2\pi}\sin\phi\,d\phi - \int_0^{\pi}\cos\theta\ \sin^2\theta\,d\theta\int_0^{2\pi}\sin\phi\,d\phi\right] = 0$$

because in both integrals $\int_0^{2\pi}\sin\phi\,d\phi = [-\cos\phi]_0^{2\pi} = -1 - (-1) = 0.$

$$\hat{L}_x^2 Y_{20} = \hat{L}_x \hat{L}_x Y_{20} = -i\hbar\left(-\sin\phi\frac{\partial}{\partial\theta} - \frac{\cos\phi}{\tan\theta}\frac{\partial}{\partial\phi}\right)\left(-i\hbar\frac{3}{2}\sqrt{\frac{5}{\pi}}\sin\phi\cos\theta\sin\theta\right)$$

$$= \hbar^2\frac{3}{2}\sqrt{\frac{5}{\pi}}\left((\cos^2\theta - \sin^2\theta)\sin^2\phi + \cos^2\theta\cos^2\phi\right)$$

$$= -\hbar^2\frac{3}{2}\sqrt{\frac{5}{\pi}}\left(\sin^2\theta\sin^2\phi - \cos^2\theta\right).$$

$$\left\langle Y_{20}\middle|\hat{L}_x^2 Y_{20}\right\rangle =$$

$$\int_0^{2\pi}\int_0^{\pi}\left(\frac{1}{4}\sqrt{\frac{5}{\pi}}(3\cos^2\theta - 1)\right)\left(-\hbar^2\frac{3}{2}\sqrt{\frac{5}{\pi}}\right)(\sin^2\theta\sin^2\phi - \cos^2\theta)\sin\theta\, d\theta\, d\phi$$

$$= \frac{-\hbar^2 15}{8\pi}\int_0^{2\pi}\int_0^{\pi}\Big[(3\cos^2\theta\sin^3\theta\sin^2\phi) - (3\cos^4\theta\sin\theta)$$

$$-(\sin^3\theta\sin^2\phi) + (\cos^2\theta\sin\theta)\Big]d\theta\, d\phi$$

$$= \frac{-\hbar^2 15}{8\pi}\left(\frac{-24}{15}\right)(\pi) = 3\hbar^2.$$

Now that we know both $\langle\hat{L}_x\rangle$ and $\langle\hat{L}_x^2\rangle$, we can calculate the uncertainty,

$$\Delta\hat{L}_x = (\langle\hat{L}_x^2\rangle - \langle\hat{L}_x\rangle^2)^{1/2} = 3^{1/2}\hbar.$$

So even though the expectation value of $\hat{L}_x = 0$, the uncertainty is a nonzero value. The same result will hold for $\Delta\hat{L}_y$. In contrast, since Y_{20} is an eigenfunction of $\hat{L}_z$, $\Delta\hat{L}_z = 0$.

26. Equation [8-60] expresses the orthogonality of the spherical harmonics. A set of functions are orthogonal if the integral of the product of any two functions is identically zero, unless it is the square of one of the functions. Thus, what must be shown here is merely that the integral over Y_{21} and Y_{11} is equal to zero. Therefore, we can ignore the values of any constants, and examine only the portions of the functions that depend on θ and ϕ.

$$Y_{21} \sim \sin\theta\cos\theta\, e^{i\phi}$$

$$Y_{11} \sim \sin\theta\, e^{i\theta}$$

$$\left\langle Y_{21} \mid Y_{11} \right\rangle \propto \int_0^{\pi} (\sin^2\theta \ \cos\theta) \sin\theta \, d\theta \int_0^{2\pi} (e^{i\phi})^* (e^{i\phi}) \, d\phi .$$

We can evaluate the integral over θ by making the substitution $u = \sin\theta$.

$$\int_0^{\pi} \sin^3\theta \ \cos\theta \, d\theta = \int_0^0 u^3 \, du = \left(\frac{u^4}{4} \right)_0^0 = 0 .$$

27. If we apply the same arguments as made in Sec. 8.6, the maximum value of J for two j = 1/2 source of angular momenta = 1/2 + 1/2 = 1. The minimum value of J is 1/2 – 1/2 = 0, so two sources of j = 1/2 coupled together result in two coupled states, $J_{12} = 0,1$.

    Coupling these states to another source of j = 1/2 can be accomplished analogously, only this source must be coupled to both the $J_{12} = 1$ and the $J_{12} = 0$. From coupling to the $J_{12} = 1$ state, the possibilities are 1 + 1/2 and 1 – 1/2, J_{123} = 3/2 and 1/2. Coupling to the $J_{12} = 0$ state leads to a single coupled state, J_{123} = 1/2. So coupling three sources results in total angular momentum states = 3/2, 1/2, and 1/2 (Note: All the possibilities are distinct, so both of the 1/2 states must be included).

    Finally, if we couple all these states to a fourth source of j = 1/2, the possible values of J_{tot} from J_{123} = 3/2 are 3/2 + 1/2 = 2 and 3/2 – 1/2 = 1. For J_{123} = 1/2, the possible J_{tot} values are 1/2 + 1/2 = 1 and 1/2 – 1/2 = 0 (for both J = 1/2 states). The total set of coupled states is J_{tot} = 2, 1, 1, 1, 0, 0. To check, compare the number of uncoupled to coupled states. The number of uncoupled states is $2 \times 2 \times 2 \times 2 = 16$. The number of coupled states is $(2 \times 2 + 1) + 3 \times (2 \times 1 + 1) + 2 \times (2 \times 0 + 1) = 5 + 3 \times 3 + 2 \times 1 = 16$. The number of states match, as they must.

28. The first step is to determine the value of the normalization constant, N, in terms of the variational parameter α.

$$1 = \langle \Gamma(x) | \Gamma(x) \rangle = N^2 \int_{-\infty}^{\infty} x^2 e^{-2\alpha x^2} dx = N^2 \frac{1}{4\alpha}\sqrt{\frac{\pi}{2\alpha}}$$

$$N = 2\sqrt{\alpha}\left(\frac{2\alpha}{\pi}\right)^{1/4}.$$

The next step is to evaluate the expectation value of the Hamiltonian with this wavefunction.

$$W = \langle \Gamma(x) | \hat{H}\, \Gamma(x) \rangle = N^2 \int_{-\infty}^{\infty} xe^{-\alpha x^2}\left(\frac{-\hbar^2}{2m}\frac{d^2}{dx^2} + \frac{1}{2}kx^2\right) xe^{-\alpha x^2} dx$$

$$= N^2 \int_{-\infty}^{\infty} xe^{-\alpha x^2}\left(\frac{3\alpha\hbar^2 x}{m} - \frac{2\alpha^2\hbar^2 x^3}{m} + \frac{1}{2}kx^3\right) e^{-\alpha x^2} dx$$

$$= N^2 \int_{-\infty}^{\infty} \left(\frac{3\alpha\hbar^2 x^2}{m} - \frac{2\alpha^2\hbar^2 x^4}{m} + \frac{1}{2}kx^4\right) e^{-2\alpha x^2} dx$$

$$= 4\alpha\sqrt{\frac{2\alpha}{\pi}}\left(\frac{3\alpha\hbar^2}{m}\frac{1}{2}\sqrt{\frac{\pi}{(2\alpha)^3}} - \frac{2\alpha^2\hbar^2}{m}\frac{3}{4}\sqrt{\frac{\pi}{(2\alpha)^5}} + \left(\frac{k}{2}\right)\frac{3}{4}\sqrt{\frac{\pi}{(2\alpha)^5}}\right)$$

$$= \frac{3\alpha\hbar^2}{2m} + \frac{3k}{8\alpha}.$$

This expression, the expectation value of the Hamiltonian operator, corresponds to the energy of the state described by this trial wavefunction.

The final step is to minimize this energy, by varying α; the value of α that minimizes the energy characterizes the best trial wavefunction.

$$\frac{dW}{d\alpha} = \frac{d}{d\alpha}\left(\frac{3\alpha\hbar^2}{2m} + \frac{3k}{8\alpha}\right) = \frac{3\hbar^2}{2m} - \frac{3k}{8\alpha^2} = 0 \Rightarrow \alpha = \frac{\sqrt{km}}{2\hbar}.$$

Now, we substitute this value of α back into the energy expression to find the minimum value of the energy.

$$E = \frac{3\hbar^2\sqrt{km}}{2m(2\hbar)} + \frac{3k(2\hbar)}{8\sqrt{km}} = \frac{3}{2}\hbar\sqrt{\frac{k}{m}} = \frac{3}{2}\hbar\omega.$$

This energy is exactly $\hbar\omega$ above the ground state; it is the exact energy of the first excited state. Thus, this trial wavefunction is the correct form of the exact $n = 1$ harmonic oscillator wavefunction.

29. Notice that the two wavefunctions being combined are proportional to the first and second ($n = 1$ and $n = 2$) harmonic oscillator functions; in other words, the trial function is a linear combination of two exact wavefunctions. To simplify the analysis, we rewrite the trial wavefunction as

$$\phi = c_1\Psi_1 + c_2\Psi_2$$

Then, the normalization of ϕ leads to the following condition on c_1 and c_2:

$$\langle\phi|\phi\rangle = \langle c_1\Psi_1 + c_2\Psi_2|\, c_1\Psi_1 + c_2\Psi_2\rangle = c_1^2\,\langle\Psi_1|\Psi_1\rangle + c_2^2\,\langle\Psi_2|\Psi_2\rangle$$

$$= c_1^2 + c_2^2 = 1.$$

Therefore, $c_2^2 = 1 - c_1^2$.

In this first step, the orthogonality of Ψ_1 and Ψ_2 ($\langle\Psi_1|\Psi_2\rangle = 0$) was used to eliminate the cross terms. Now, when we evaluate the expectation value of the energy, we use this relationship to reduce the number of variables to one.

$$\langle\phi|\hat{H}\,\phi\rangle = \langle c_1\Psi_1 + c_2\Psi_2|\,\hat{H}\,|c_1\Psi_1 + c_2\Psi_2\rangle = c_1^2\,E_1 + (1 - c_1^2)\,E_2$$

$$= c_1^2\,(E_1 - E_2) + E_2.$$

Because c_1^2 is a monotonically increasing function that can vary from 0 to 1, by inspection we see that the lowest value of the energy is obtained when $c_1^2 = 1.0$. In other words, the best trial wavefunction that is a combination of exact wavefunctions is just the exact wavefunction with the lowest energy.

30. We follow the same approach as in Problem 28, but we replace $1/2\ kx^2$ with $1/2\ kx^4$.

$$\left\langle \Gamma(x)|\hat{H}\,\Gamma(x)\right\rangle = N^2 \int_{-\infty}^{\infty} xe^{-\alpha x^2}\left(\frac{-\hbar^2}{2m}\frac{d^2}{dx^2}+\frac{1}{2}kx^4\right)xe^{-\alpha x^2}\,dx$$

$$= N^2 \int_{-\infty}^{\infty} xe^{-\alpha x^2}\left(\frac{3\alpha\hbar^2 x}{m}-\frac{2\alpha^2\hbar^2 x^3}{m}+\frac{1}{2}kx^5\right)e^{-\alpha x^2}\,dx$$

$$= N^2 \int_{-\infty}^{\infty}\left(\frac{3\alpha\hbar^2 x^2}{m}-\frac{2\alpha^2\hbar^2 x^4}{m}+\frac{1}{2}kx^6\right)e^{-2\alpha x^2}\,dx$$

$$= 4\alpha\sqrt{\frac{2\alpha}{\pi}}\left(\frac{3\alpha\hbar^2}{m}\frac{1}{2}\sqrt{\frac{\pi}{(2\alpha)^3}}-\frac{2\alpha^2\hbar^2}{m}\frac{3}{4}\sqrt{\frac{\pi}{(2\alpha)^5}}+\frac{k}{2}\frac{15}{8}\sqrt{\frac{\pi}{(2\alpha)^7}}\right)$$

$$= \frac{3\alpha\hbar^2}{2m}+\frac{15k}{32\alpha^2}.$$

Now, we minimize the expectation value of the Hamiltonian operator to determine the value of α that minimizes the energy:

$$\frac{dW}{d\alpha}=\frac{d}{d\alpha}\left(\frac{3\alpha\hbar^2}{2m}+\frac{15k}{32\alpha^2}\right)=\frac{3\hbar^2}{2m}-\frac{30k}{32\alpha^3}=0\Rightarrow\alpha=\left(\frac{5km}{8\hbar^2}\right)^{1/3}.$$

We substitute this value of α back into the energy expression to find the minimum value of the energy:

$$E=\frac{3\hbar^2}{2m}\left(\frac{5km}{8\hbar^2}\right)^{1/3}+\frac{15k}{32}\left(\frac{8\hbar^2}{5km}\right)^{2/3}=\frac{9}{8}\left(\frac{5k\hbar^4}{m^2}\right)^{1/3}.$$

31. The first-order correction to the energy is simply the expectation value of the perturbing Hamiltonian [8-73]:

$$\langle \Psi_n^{(0)}|\hat{H}^{(1)}|\Psi_n^{(0)}\rangle = E_n^{(1)}.$$

With $\hat{H}^{(1)} = gx^4$, the first-order correction to the ground-state energy is

$$E_0^{(1)} = \int_{-\infty}^{\infty}\left(\frac{\beta^2}{\pi}\right)^{1/4} e^{-\beta^2x^2/2}\, gx^4 \left(\frac{\beta^2}{\pi}\right)^{1/4} e^{-\beta^2x^2/2}\, dx.$$

$$= g\left(\frac{\beta^2}{\pi}\right)^{1/2} \int_{-\infty}^{\infty} x^4 e^{-\beta^2x^2}\, dx = g\left(\frac{\beta^2}{\pi}\right)^{1/2} \frac{3}{4}\left(\frac{\pi}{\beta^{10}}\right)^{1/2} = \frac{3g}{4\beta^4}.$$

For the first excited state:

$$E_1^{(1)} = \int_{-\infty}^{\infty}\left(\frac{\beta^2}{4\pi}\right)^{1/4} 2\beta x\, e^{-\beta^2x^2/2}\, gx^4 \left(\frac{\beta^2}{4\pi}\right)^{1/4} 2\beta x\, e^{-\beta^2x^2/2}\, dx$$

$$= 4\beta^2 g\left(\frac{\beta^2}{4\pi}\right)^{1/2} \int_{-\infty}^{\infty} x^6 e^{-\beta^2x^2}\, dx = \frac{2g\beta^3}{\sqrt{\pi}}\,\frac{15}{8}\left(\frac{\pi}{\beta^{14}}\right)^{1/2} = \frac{15g}{4\beta^4}.$$

For the second excited state:

$$E_2^{(1)} = \int_{-\infty}^{\infty}\left(\frac{\beta^2}{64\pi}\right)^{1/4} \left(4\beta^2x^2 - 2\right) e^{-\beta^2x^2/2}\, gx^4 \left(\frac{\beta^2}{64\pi}\right)^{1/4} \left(4\beta^2x^2 - 2\right) e^{-\beta^2x^2/2}\, dx$$

$$= g\left(\frac{\beta^2}{64\pi}\right)^{1/2} \int_{-\infty}^{\infty} \left(16\beta^4x^8 - 16\beta^2x^6 + 4x^4\right) e^{-\beta^2x^2}\, dx$$

$$= \frac{g\beta}{8\sqrt{\pi}}\left[\left(16\beta^4\right)\frac{105}{16}\left(\frac{\pi}{\beta^{18}}\right)^{1/2} - \left(16\beta^2\right)\frac{15}{8}\left(\frac{\pi}{\beta^{14}}\right)^{1/2} + (4)\frac{3}{4}\left(\frac{\pi}{\beta^{10}}\right)^{1/2}\right]$$

$$= \frac{g\beta}{8\sqrt{\pi}}(105 - 30 + 3)\frac{\sqrt{\pi}}{\beta^5} = \frac{39g}{4\beta^4}.$$

32. With $\hat{H}^{(1)} = gx^3$, the first-order correction to the ground-state energy is

$$E_0^{(1)} = \int_{-\infty}^{\infty}\left(\frac{\beta^2}{\pi}\right)^{1/4} e^{-\beta^2x^2/2}\, gx^3 \left(\frac{\beta^2}{\pi}\right)^{1/4} e^{-\beta^2x^2/2}\, dx = 0.$$

The first-order correction for $gx^3 = 0$ because x^3 odd, and the ground-state wavefunction is even (symmetric about $x = 0$).

To determine the first-order correction to the wavefunction requires the application of [8-77] to determine the coefficients, c_i:

$$c_i = \frac{\langle \Psi_i^{(0)} | \hat{H}^{(1)} | \Psi_0^{(0)} \rangle}{E_0^{(0)} - E_i^{(0)}}.$$

Based on the symmetry of the perturbing Hamiltonian, the only states for which the numerator in the preceding expression can be nonzero are for odd values of i; then the product of the odd state i, the odd perturbing Hamiltonian, and the even ground-state wavefunction are overall even.

$$\langle \Psi_1^{(0)} | \hat{H}^{(1)} | \Psi_0^{(0)} \rangle = \int_{-\infty}^{\infty} \left(\frac{\beta^2}{4\pi}\right)^{1/4} 2\beta x\, e^{-\beta^2 x^2/2}\, gx^3 \left(\frac{\beta^2}{\pi}\right)^{1/4} e^{-\beta^2 x^2/2}\, dx$$

$$= 2\beta g \left(\frac{\beta^2}{2\pi}\right)^{1/2} \int_{-\infty}^{\infty} x^4 e^{-\beta^2 x^2}\, dx$$

$$= 2\beta g \left(\frac{\beta^2}{2\pi}\right)^{1/2} \frac{3}{4}\sqrt{\frac{\pi}{\beta^{10}}} = \frac{3g}{(\sqrt{2}\,\beta)^3}.$$

$$c_1 = \frac{\langle \Psi_1^{(0)} | \hat{H}^{(1)} | \Psi_0^{(0)} \rangle}{E_0^{(0)} - E_1^{(0)}} = \frac{3g}{(\sqrt{2}\,\beta)^3\left(\frac{1}{2}\hbar\omega - \frac{3}{2}\hbar\omega\right)} = -\frac{3g}{(\sqrt{2}\,\beta)^3\,\hbar\omega}.$$

$$\langle \Psi_3^{(0)} | \hat{H}^{(1)} | \Psi_0^{(0)} \rangle$$

$$= \int_{-\infty}^{\infty} \left(\frac{\beta^2}{2304\pi}\right)^{1/4} (8(\beta x)^3 - 12\beta x)\, e^{-\beta^2 x^2/2}\, gx^3 \left(\frac{\beta^2}{\pi}\right)^{1/4} e^{-\beta^2 x^2/2}\, dx$$

$$= \frac{4\beta^2 g}{2\sqrt{12}\,\pi}\left[\int_{-\infty}^{\infty} 2\beta^2 x^6 e^{-\beta^2 x^2}\, dx - \int_{-\infty}^{\infty} 3x^4 e^{-\beta^2 x^2}\, dx\right]$$

$$= \frac{2\beta^2 g}{\sqrt{12\,\pi}}\left[2\beta^2 \frac{15}{8}\sqrt{\frac{\pi}{\beta^{14}}} - (3)\frac{3}{4}\sqrt{\frac{\pi}{\beta^{10}}}\right] = \frac{\sqrt{3}\,g}{2\,\beta^3}.$$

$$c_3 = \frac{\left\langle \Psi_3^{(0)} | \hat{H}^{(1)} | \Psi_0^{(0)} \right\rangle}{E_0^{(0)} - E_3^{(0)}} = \frac{\sqrt{3}\,g}{2\beta^3\left(\frac{1}{2}\hbar\omega - \frac{7}{2}\hbar\omega\right)} = -\frac{g}{2\sqrt{3}\,\beta^3\hbar\omega}.$$

It turns out that these are the only nonzero coefficients (you may want to try c_5 just to see that this is the case). Therefore, we can write the wavefunction to first order as

$$\Psi \approx \Psi_0^{(0)} - \frac{3g}{(\sqrt{2}\,\beta)^3\,\hbar\omega}\Psi_1^{(0)} - \frac{g}{2\sqrt{3}\,\beta^3\hbar\omega}\Psi_3^{(0)}.$$

With the first-order correction to the wavefunction known, we can find the second-order correction to the energy using [8-74]:

$$E_0^{(2)} = \left\langle \Psi_0^{(0)} | (\hat{H}^{(1)} - E_0^{(0)}) | \Psi_0^{(1)} \right\rangle = \left\langle \Psi_0^{(0)} | (\hat{H}^{(1)} - E_0^{(0)}) | c_1\Psi_1^{(0)} + c_3\Psi_3^{(0)} \right\rangle.$$

$$= c_1\left\langle \Psi_0^{(0)} | \hat{H}^{(1)} | \Psi_1^{(0)} \right\rangle + c_3\left\langle \Psi_0^{(0)} | \hat{H}^{(1)} | \Psi_3^{(0)} \right\rangle$$

$$= -\frac{3g}{(\sqrt{2}\,\beta)^3\,\hbar\omega}\left(\frac{3g}{(\sqrt{2}\,\beta)^3}\right) - \frac{g}{2\sqrt{3}\,\beta^3\hbar\omega}\left(\frac{\sqrt{3}\,g}{2\beta^3}\right) = -\frac{11g^2}{8\,\beta^6\hbar\omega}.$$

33. To derive the general expression for the third-order correction to the energy, we follow the same approach used for the first and second. First, we collect all the terms that depend on the perturbation parameter, λ, to third order (λ^3):

$$\hat{H}^{(3)}\Psi^{(0)} + \hat{H}^{(2)}\Psi^{(1)} + \hat{H}^{(1)}\Psi^{(2)} + \hat{H}^{(0)}\Psi^{(3)} =$$

$$E^{(3)}\Psi^{(0)} + E^{(2)}\Psi^{(1)} + E^{(1)}\Psi^{(2)} + E^{(0)}\Psi^{(3)}$$

Now, group the Hamiltonian and energy terms together.

$$(\hat{H}^{(3)} - E^{(3)})\Psi^{(0)} + (\hat{H}^{(2)} - E^{(2)})\Psi^{(1)} + (\hat{H}^{(1)} - E^{(1)})\Psi^{(2)} + (\hat{H}^{(0)} - E^{(0)})\Psi^{(3)} = 0.$$

We multiply this entire equation by the zeroth-order wavefunction and integrate. The last term will go to zero as explained on the bottom of p. 323 in the text.

$$\langle\Psi^{(0)}|\,(\hat{H}^{(3)} - E^{(3)})\Psi^{(0)}\rangle + \langle\Psi^{(0)}|\,(\hat{H}^{(2)} - E^{(2)})\Psi^{(1)}\rangle + \langle\Psi^{(0)}|\,(\hat{H}^{(1)} - E^{(1)})\Psi^{(2)}\rangle = 0.$$

In the first term, $E^{(3)}$ is just a constant, and so that term reduces to an overlap integral of the zeroth-order wavefunction (which equals 1), and so finally we can write the general expression for the third-order energy correction:

$$\langle\Psi^{(0)}|\,\hat{H}^{(3)}\,\Psi^{(0)}\rangle + \langle\Psi^{(0)}|\,(\hat{H}^{(2)} - E^{(2)})\Psi^{(1)}\rangle + \langle\Psi^{(0)}|\,(\hat{H}^{(1)} - E^{(1)})\Psi^{(2)}\rangle = E^{(3)}.$$

Note that this expression requires the second-order correction to the wavefunction. It is possible with further manipulation to eliminate the second-order correction to the wavefunction and then the third-order correction to the energy can be obtained with just the zeroth- and first-order corrections to the wavefunction.

34. As in the previous problem, we group all the terms from the normalization equation that depend on the perturbation parameter to second order:

$$\langle\Psi^{(2)}|\Psi^{(0)}\rangle + \langle\Psi^{(1)}|\Psi^{(1)}\rangle + \langle\Psi^{(0)}|\Psi^{(2)}\rangle = 0.$$

35. For the one-dimensional particle in a box of length l, the wavefunctions are

$$\Psi_n(x) = \sqrt{\frac{2}{l}}\,\sin\left(\frac{n\pi x}{l}\right).$$

The first-order corrections to the energy are given by

$$E_n^{(1)} = \int_0^l \sqrt{\frac{2}{l}}\,\sin\left(\frac{n\pi x}{l}\right)\,ax\,\sqrt{\frac{2}{l}}\,\sin\left(\frac{n\pi x}{l}\right)dx$$

$$= \frac{2a}{l}\int_0^l x\sin^2\left(\frac{n\pi x}{l}\right)dx$$

$$= \frac{2a}{l}\left(\frac{x^2}{4} - \frac{x\sin(2n\pi x/l)}{4n\pi x/l} - \frac{\cos(2n\pi x/l)}{8(n\pi x/l)^2}\right)_0^l = \frac{a}{2l}.$$

For this problem, the first-order correction to the energy is identical for all states.

The first-order correction to the wavefunction is a linear combination of the unperturbed wavefunctions, and each coefficient must be determined. From [8-77], the coefficient for the first-order correction to the nth state is

$$c_j = \left[\int_0^l \sqrt{\frac{2}{l}}\sin\left(\frac{j\pi x}{l}\right) ax \sqrt{\frac{2}{l}}\sin\left(\frac{n\pi x}{l}\right) dx\right] \Big/ \left(\frac{(n^2 - j^2)h^2}{8\,m\,l^2}\right)$$

$$= \frac{16\,m\,a\,l}{h^2(n^2 - j^2)}\int_0^l x\,\sin\left(\frac{j\pi x}{l}\right)\sin\left(\frac{n\pi x}{l}\right) dx.$$

This integral can be found in a mathematical handbook.

$$\int_0^l x\,\sin\left(\frac{j\pi x}{l}\right)\sin\left(\frac{n\pi x}{l}\right) dx = \frac{-4nj}{\pi^2(n-j)^2(n+j)^2}, \text{ if } n+j \text{ is odd.}$$

Substitution of this result back into the previous expression leads to

$$c_j = \frac{-64nj\,m\,a\,l^3}{(\hbar\pi)^2(n-j)^3(n+j)^3}, \text{ if } n+j \text{ is odd.}$$

Then the first order correction to the wavefunction is written as

$$\Psi_n^{(1)} = \sum_j \frac{-64n\,j\,m\,a\,l^3}{(\hbar\pi)^2(n-j)^3(n+j)^3}\Psi_j^{(0)}, \text{ where } j+n \text{ is odd.}$$

Now, we can substitute this result into the expression for the second-order energy expression to determine $E^{(2)}$.

$$E_n^{(2)} = \left\langle \Psi_n^{(0)} | \hat{H}^{(1)} | \Psi_n^{(1)} \right\rangle. \qquad \text{(see [8-74])}$$

(In [8-74], the first term is zero because there is no $H^{(2)}$, and likewise, because $\Psi^{(0)}$ and $\Psi^{(1)}$ are orthogonal, the term containing $E^{(1)}$ is zero.)

$$E_n^{(2)} = \left\langle \Psi_n^{(0)} | \hat{H}^{(1)} | \sum_j c_j \Psi_j^{(0)} \right\rangle = \sum_j c_j \langle \Psi_n^{(0)} | \hat{H}^{(1)} | \Psi_j^{(0)} \rangle$$

$$= \sum_j \frac{\langle \Psi_n^{(0)} | \hat{H}^{(1)} | \Psi_j^{(0)} \rangle^2}{E_n^{(0)} - E_j^{(0)}} = \sum_j \frac{512(nja)^2 ml^4}{\pi^4 h^2 (n-j)^5 (n+j)^5}$$

The substitution in the last step used the results from the calculation of c_j.

36. To make a qualitative comparison between the true particle in a box ($V(x) = 0$ for $0 < x < l$) and the potential in this problem, where $V(x) = V_0$ for $x_a < x < x_b$, it is helpful to rewrite the Schrödinger equation as

$$(d/dx)^2 \Psi = 2m (E - V(x)) \Psi / \hbar.$$

In this form, it is clear that when $E > V(x)$, then the second derivative (also called the *curvature*) will be positive if the wavefunction is positive (and negative if the wavefunction is negative). A function with a positive curvature is concave up; a negative curvature indicates the function will be concave down.

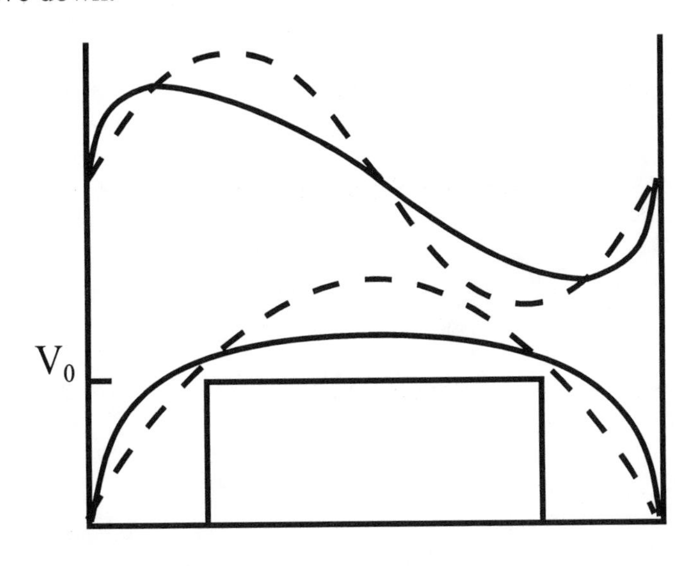

Here, the discontinuity of the potential (jumping from zero → V_0) means that the second derivative must be discontinuous as well. If $E = V_0$, then the second derivative would be zero in this region; the wavefunction would then have to be linear. In the specific case of this problem, the second derivative will be small in this region of the potential.

Note how the perturbed (solid lines) wavefunctions are "flattened out" in the region of the potential bump compared to the unperturbed particle in a box wavefunctions (broken lines).

37. The first-order energy correction is given by

$$E_n^{(1)} = \int_0^l \sqrt{\frac{2}{l}} \sin\left(\frac{n\pi x}{l}\right)\left(0.1\sqrt{\frac{2}{l}} \sin\left(\frac{\pi x}{l}\right)\right)\sqrt{\frac{2}{l}} \sin\left(\frac{n\pi x}{l}\right) dx$$

$$= 0.1\left(\frac{2}{l}\right)^{3/2} \int_0^l \sin^2\left(\frac{n\pi x}{l}\right) \sin\left(\frac{\pi x}{l}\right) dx .$$

The integral can be looked up in a table of integrals. The final result is

$$E_n^{(1)} = 0.1\left(\frac{2}{l}\right)^{3/2}\left(\frac{l}{\pi}\right)\left(\frac{4n^2}{4n^2 - 1}\right).$$

Chapter 9

Vibrational-Rotational Spectroscopy

Exercises

1. To convert wavenumbers (cm^{-1}) to frequency (Hz), recall the relationship between the frequency and wavelength of electromagnetic radiation, $\nu\lambda = c$, which implies that $\nu = c \times (1/\lambda)$, and since $1/\lambda$ is in units of cm^{-1}, the value to use for c is 2.998×10^{10} cm s^{-1}. To convert wavenumbers to J mol^{-1}, we use the conversion factor in Appendix VI:

$$1\ cm^{-1} \times (1.986 \times 10^{-23}\ J / cm^{-1}) \times (6.022 \times 10^{23}\ mol^{-1})$$
$$= 11.96\ (J\ mol^{-1}/cm^{-1}).$$

1 cm^{-1}	2.998×10^{10} Hz	11.96 J mol^{-1}	microwave
10 cm^{-1}	2.998×10^{11} Hz	119.6 J mol^{-1}	infrared
100 cm^{-1}	2.998×10^{12} Hz	1196 J mol^{-1}	infrared
1000 cm^{-1}	2.998×10^{13} Hz	11,960 J mol^{-1}	infrared
10,000 cm^{-1}	2.998×10^{14} Hz	119,600 J mol^{-1}	visible

2. Use [9–48] to find the rotational constant.

$$\tilde{B}_e = \frac{0.01\,h}{8\pi^2 \mu r_e^2 c} = \frac{(0.01 \text{ m cm}^{-1})(6.626 \times 10^{-34} \text{ J s})}{(8\pi^2)(2.998 \times 10^8 \text{ m s}^{-1})\mu r_e^2}$$
$$= \frac{2.799 \times 10^{-46} \text{ cm}^{-1} \text{ kg m}^2}{\mu r_e^2}.$$

$r_e = 1.595$ Å $= 1.595 \times 10^{-10}$ m.

masses: H = 1.00785 amu, D = 2.01410 amu, ^{7}Li = 7.01600, and ^{6}Li = 6.01512 amu.

$\mu_{\text{LiH}} = (1.00785)(7.01600) / (1.00785 + 7.01600) = 0.88126$ amu.

$$\tilde{B}_e(\text{LiH}) = \frac{2.799 \times 10^{-46} \text{ cm}^{-1} \text{ kg m}^2}{(0.88126 \text{ amu})(1.6606 \times 10^{-27} \text{ kg amu}^{-1})(1.595 \times 10^{-10} \text{ m})^2}$$
$$= 7.519 \text{ cm}^{-1}.$$

$\mu_{\text{LiD}} = (2.01410)(7.01600) / (2.01410 + 7.01600) = 1.5648$ amu.

$$\tilde{B}_e(\text{LiD}) = \frac{2.799 \times 10^{-46} \text{ cm}^{-1} \text{ kg m}^2}{(1.5648 \text{ amu})(1.6606 \times 10^{-27} \text{ kg amu}^{-1})(1.595 \times 10^{-10} \text{ m})^2}$$
$$= 4.234 \text{ cm}^{-1}.$$

$\mu_{^6\text{LiH}} = (1.00785)(6.01512) / (1.00785 + 6.01512) = 0.8632$ amu.

$$\tilde{B}_e(^6\text{LiH}) = \frac{2.799 \times 10^{-46} \text{ cm}^{-1} \text{ kg m}^2}{(0.86322 \text{ amu})(1.6606 \times 10^{-27} \text{ kg amu}^{-1})(1.595 \times 10^{-10} \text{ m})^2}$$
$$= 7.677 \text{ cm}^{-1}.$$

3. For any given n → n + 1 transition, the first line in the R–branch is the n,0 → n+1,1 transition. The first line in the P–branch is the n,1 → n+1,0 transition. We can calculate the energy of these transitions by using the energy expression at the bottom of p. 375, where the only terms included are those for which constants are given in Table 9.1.

$$E_{n+1,1} = \widetilde{\omega}_e[(n+1)+1/2] + \widetilde{B}_e(1)(2) - \widetilde{\omega}_e\widetilde{\chi}_e[(n+1)+1/2]^2$$
$$- \widetilde{\alpha}_e(1)(2)[(n+1)+1/2].$$

$$E_{n,0} = \widetilde{\omega}_e(n+1/2) + \widetilde{B}_e(0)(1) - \widetilde{\omega}_e\widetilde{\chi}_e(n+1/2)^2 - \widetilde{\alpha}_e(0)(1)(n+1/2).$$

$$\Delta E_{n,0\rightarrow n+1,1} = E_{n+1,1} - E_{n,o} = \widetilde{\omega}_e + 2(n+1)\widetilde{\omega}_e\widetilde{\chi}_e + 2\widetilde{B}_e - \widetilde{\alpha}_e(2n+3).$$

$$E_{n+1,0} = \widetilde{\omega}_e([n+1]+1/2) + \widetilde{B}_e(0)(1) - \widetilde{\omega}_e\widetilde{\chi}_e([n+1]+1/2)^2$$
$$- \widetilde{\alpha}_e(0)(1)([n+1]+1/2).$$

$$E_{n,1} = \widetilde{\omega}_e(n+1/2) + \widetilde{B}_e(1)(2) - \widetilde{\omega}_e\widetilde{\chi}_e(n+1/2)^2 - \widetilde{\alpha}_e(1)(2)(n+1/2).$$

$$\Delta E_{n,1\rightarrow n+1,0} = E_{n+1,0} - E_{n,1} = \widetilde{\omega}_e - 2(n+1)\widetilde{\omega}_e\widetilde{\chi}_e - 2\widetilde{B}_e + \widetilde{\alpha}_e(2n+1).$$

$$(\Delta E_{n,0\rightarrow n+1,1} + \Delta E_{n,1\rightarrow n+1,0})/2 = \widetilde{\omega}_e - 2(n+1)\widetilde{\omega}_e\widetilde{\chi}_e - \widetilde{\alpha}_e.$$

Using this expression, we can now calculate the band center for the fundamental and overtone transitions. From Table 9.1, for LiH

$$\widetilde{\omega}_e = 1405.6,\ \widetilde{\omega}_e\widetilde{\chi}_e = 23.20\ \text{cm}^{-1}, \text{and } \widetilde{\alpha}_e = 0.2132\ \text{cm}^{-1}.$$

$$n = 0\rightarrow 1:\quad 1405.6 - (2)(23.20) - 0.2132 = 1359.0\ \text{cm}^{-1}.$$

$$n = 1\rightarrow 2:\quad 1405.6 - (4)(23.20) - 0.2132 = 1312.6\ \text{cm}^{-1}.$$

$$n = 2\rightarrow 3:\quad 1405.6 - (6)(23.20) - 0.2132 = 1266.2\ \text{cm}^{-1}.$$

$$n = 3\rightarrow 4:\quad 1405.6 - (8)(23.20) - 0.2132 = 1219.8\ \text{cm}^{-1}.$$

4. The standard value of the bond length of a C≡C bond is 1.20 Å, of a C–H bond is 1.10Å, and of a C–F bond is 1.28 Å. The masses are H = 1.00783 amu, C = 12.000 amu, and F = 18.9984 amu. We use [9–53] to calculate the moments of inertia.

$$I(\text{HCCH}) = \Big\{(2)(1.00783)(12.000)[(1.10)^2 + (2.30)^2] + (1.00783)^2(3.40)^2$$
$$+(12.000)^2(1.20)^2\Big\} / [(2)(1.00783 + 12.000)]$$

$= 14.373 \text{ amu Å}^2.$

$= (14.373 \text{ amu Å}^2)(1.661 \times 10^{-27} \text{ kg amu}^{-1})(10^{-10} \text{ m Å}^{-1})^2$

$= 2.387 \times 10^{-46} \text{ kg m}^2.$

To express the rotational constant in terms of megahertz (MHz), [9–55] relates the energy to the moment of inertia, and by equating the energy to hν,

$$E = J(J+1)\frac{\hbar^2}{2I} = J(J+1)B_e, \text{ so } B_e = \frac{\hbar^2}{2I}.$$

$$h\nu = J(J+1)B_e(J); \; B_e / h = \hbar / 4\pi I \text{ [Hz]}.$$

$$B(HCCH) = \frac{1.0545 \times 10^{-34} \text{ J s}}{(4\pi)2.387 \times 10^{-46} \text{ kg m}^2} = 3.516 \times 10^{10} \text{ Hz}$$

$$= 35{,}160 \text{ MHz} = 35.16 \text{ GHz}.$$

$$\begin{aligned} I(FCCH) = {} & [(18.9984)(12.000)(1.28^2 + 2.48^2) \\ & + (18.9984)(1.00783)(3.58)^2 + (12.000)^2(1.2)^2 \\ & + (1.00783)(12.000)([1.1]^2 + [2.3]^2)] \\ & \div [(2)(12.000) + 18.9984 + 1.00783] \end{aligned}$$

$= 61.566 \text{ amu Å}^2$

$= (61.566 \text{ amu Å}^2)(1.661 \times 10^{-27} \text{ kg amu}^{-1})(10^{-10} \text{ m Å}^{-1})^2$

$= 1.023 \times 10^{-45} \text{ kg m}^2.$

$$B(FCCH) = \frac{1.0545 \times 10^{-34} \text{ J s}}{(4\pi)1.023 \times 10^{-45} \text{ kg m}^2} = 8.203 \times 10^9 \text{ Hz}$$

$$= 8203 \text{ MHz} = 8.203 \text{ GHz}.$$

5. We recall that the criterion for an infrared mode to be active is that the dipole moment change with the motion of the normal mode [p. 362, 363]. The first two modes maintain the symmetry of the equilibrium structure of acetylene; therefore, the dipole moment remains zero whatever the value of

the normal mode coordinate. Because the magnitude of the dipole moment does not change, these two modes are inactive.

The third and fourth modes pictured break the symmetry of acetylene, and once the symmetry is broken, the molecule possesses a nonzero dipole moment. These modes are active.

The last mode maintains some of the symmetry of the acetylene molecule; the symmetry that allows us to rotate the acetylene 180° about an axis perpendicular to the plane of the normal mode motion prevents the molecule from possessing a dipole moment. Thus, this last mode is inactive.

Additional Exercises

6. Since the system is changing with time, we must work with the stationary-state wavefunctions that explicitly depend on time. As shown in [9-7], they are combinations of the spatial wavefunctions we dealt with previously (ψ_0 and ψ_1) and the time-dependent phase factor, $e^{-iE_i t/\hbar}$:

$$\Psi(x,t) = a\,\psi_0 e^{-iE_0 t/\hbar} + b\,\psi_1 e^{-iE_1 t/\hbar}$$

$$= \frac{1}{\sqrt{2}}\left(\frac{\beta^2}{\pi}\right)^{1/4} e^{-\frac{\beta^2 x^2}{2}}\, e^{-i\omega t/2} + \frac{1}{\sqrt{2}}\left(\frac{\beta^2}{4\pi}\right)^{1/4} (2\beta x) e^{-\frac{\beta^2 x^2}{2}}\, e^{-3i\omega t/2}.$$

The inverse factors of $\sqrt{2}$ are required to ensure that $\Psi(x,t)$ is normalized.

The probability density at $x = 0$ is given by $\Psi(x,t)^* \Psi(x,t)$:

$$\Psi(0,t) = \frac{1}{\sqrt{2}}\psi_0(0)\, e^{-iE_0 t/\hbar} + \frac{1}{\sqrt{2}}\psi_1(0)\, e^{-iE_1 t/\hbar}$$

$$= \frac{1}{\sqrt{2}}\left(\frac{\beta^2}{\pi}\right)^{1/4} e^0\, e^{-i\omega t/2} + \frac{1}{\sqrt{2}}\left(\frac{\beta^2}{4\pi}\right)^{1/4} (2\beta(0)) e^0 e^{-3i\omega t/2}$$

$$= \frac{1}{\sqrt{2}}\left(\frac{\beta^2}{\pi}\right)^{1/4} e^{-i\omega t/2}.$$

$$\Psi(0,t) * \Psi(0,t) = \frac{1}{\sqrt{2}}\left(\frac{\beta^2}{\pi}\right)^{1/4} e^{i\omega t/2} \frac{1}{\sqrt{2}}\left(\frac{\beta^2}{\pi}\right)^{1/4} e^{-i\omega t/2}$$

$$= \frac{1}{2}\left(\frac{\beta^2}{\pi}\right)^{1/2}.$$

The probability density at x = 0 for this superposition wavefunction is constant in time. Note that it is only half of the value of the probability density of the ground-state wavefunction alone.

7. An electrically charged particle interacts with electromagnetic radiation through the electrical potential, V (unlike a dipole, which interacts with the electric field). If we assume that we have a uniform electric field, E, in the box, the associated potential at position x is $V_0 - xE$ (since the electric field is the negative of the gradient of the potential; this potential is associated with a uniform electric field); V_0 is the potential at x = 0, which we can arbitrarily choose to be zero. Thus, we can write the interaction Hamiltonian as

$$\hat{H}^{(1)} = -qxE.$$

The selection rules are determined by the integral represented by

$$\langle \psi_n \mid \hat{H}^{(1)} \psi_m \rangle.$$

When this integral is nonzero, transitions are allowed between states n and m (see [9-13]). For the particle in a box of length l

$$\langle \psi_n \mid \hat{H}^{(1)} \psi_m \rangle = \int_0^l \sqrt{\frac{2}{l}} \sin\left(\frac{n\pi x}{l}\right)(-qEx)\sqrt{\frac{2}{l}} \sin\left(\frac{m\pi x}{l}\right) dx$$

$$= \frac{-2qE}{l} \int_0^l x \sin\left(\frac{n\pi x}{l}\right) \sin\left(\frac{m\pi x}{l}\right) dx$$

$$= \frac{-2qE}{l}\left[\frac{l^2 n\, m\,(-2 \pm 2)}{(n^2 - m^2)^2 \pi^2}\right].$$

where the plus sign applies when n + m is even, and the minus sign applies when n + m is odd. Obviously, the integral is zero when n + m is even, so the selection rule is that transitions are allowed when n + m is odd.

8. The key to evaluating $\langle Y_{JM} \mid \cos\theta \; Y_{J'M'}\rangle$ is a recurrence relationship that applies to the Legendre polynomials. Many sets of orthogonal polynomials can be evaluated by recurrence; for example, see the relationship for the hermite polynomials at the bottom of p. 251 in the text.

   For the Legendre polynomials, the relationship is

   $$(J+1-|M|)\, P_{J+1}^{M}(\cos\theta) = (2J+1)\cos\theta\, P_J^M(\cos\theta) - (J+|M|)\, P_{J-1}^{M}(\cos\theta),$$

   which can be rearranged to form

   $$\cos\theta\, P_J^M(\cos\theta) = \Big((J+1-|M|)\, P_{J+1}^{M}(\cos\theta) + (J+|M|)\, P_{J-1}^{M}(\cos\theta)\Big)/(2J+1).$$

   You can verify this relationship by using it to generate two or three of the Legendre polynomials listed on p. 303.

   Since the spherical harmonics are proportional to the Legendre polynomials [8-57,59], we can now write $\cos\theta\, Y_{J'M'}$ as

   $$\cos\theta\, Y_{JM} = \cos\theta \sqrt{\frac{(2J+1)(J-|M|)!}{2(J+|M|)!}}\, P_J^M(\cos\theta)\, \frac{e^{im\phi}}{\sqrt{2\pi}}\, (-1)^{[M+|M|]/2}$$

   $$= \sqrt{\frac{(2J+1)(J-|M|)!}{2(J+|M|)!}}\, \frac{e^{im\phi}}{\sqrt{2\pi}}\, (-1)^{[M+|M|]/2} \cos\theta\, P_J^M(\cos\theta)$$

   $$= \sqrt{\frac{(2J+1)(J-|M|)!}{2(J+|M|)!}}\, \frac{e^{im\phi}}{\sqrt{2\pi}}\, (-1)^{[M+|M|]/2}\, \frac{(J+1-|M|)P_{J+1}^{M} + (J+|M|)P_{J-1}^{M}}{2J+1}$$

$$= \frac{(J+1-|M|)}{2J+1}\sqrt{\frac{(2J+1)(J-|M|)!}{2(J+|M|)!}}P_{J+1}^{M}\frac{e^{im\phi}}{\sqrt{2\pi}}(-1)^{[M+|M|]/2}$$

$$+\frac{(J+|M|)}{2J+1}\sqrt{\frac{(2J+1)(J-|M|)!}{2(J+|M|)!}}P_{J-1}^{M}\frac{e^{im\phi}}{\sqrt{2\pi}}(-1)^{[M+|M|]/2}$$

$$= \sqrt{\frac{(J+1-|M|)(J+1-|M|)}{(2J+1)(2J+3)}}\sqrt{\frac{(2J+3)(J+1-|M|)!}{2(J+1-|M|)!}}P_{J+1}^{M}\frac{e^{im\phi}}{\sqrt{2\pi}}(-1)^{[M+|M|]/2}$$

$$+\sqrt{\frac{(J+|M|)(J-|M|)}{(2J+1)(2J-1)}}\sqrt{\frac{(2J-1)(J-1-|M|)!}{2(J-1+|M|)!}}P_{J-1}^{M}\frac{e^{im\phi}}{\sqrt{2\pi}}(-1)^{[M+|M|]/2}$$

$$= \sqrt{\frac{(J+M+1)(J-M+1)}{(2J+1)(2J+3)}}Y_{J'+1,M'}+\sqrt{\frac{(J+M)(J-M)}{(2J+1)(2J-1)}}Y_{J'-1,M'}.$$

Now, we can substitute this expression into [9-17] for $\cos\theta\, Y_{J'M'}$:

$$\langle Y_{JM}|\cos\theta\, Y_{J'M'}\rangle = \sqrt{\frac{(J'+M'+1)(J-M'+1)}{(2J'+1)(2J+3)}}\langle Y_{JM}|Y_{J'+1,M'}\rangle + \sqrt{\frac{(J'+M')(J'-M')}{(2J'+1)(2J-1)}}\langle Y_{JM}|Y_{J'-1,M'}\rangle.$$

The orthogonality of the spherical harmonics reduces these integrals to zero unless J'+1 = J or J'–1 = J and M' = M.

$$\langle Y_{JM}|\cos\theta\, Y_{J'M'}\rangle = \sqrt{\frac{(J+M)(J-M)}{(2J-1)(2J-1)}}\delta_{J,J'+1} + \sqrt{\frac{(J'+M)(J'-M)}{(2J'+1)(2J'-1)}}\delta_{J,J'-1}$$

$$= \sqrt{\frac{J^2-M^2}{4J^2-1}}\delta_{J,J'+1} + \sqrt{\frac{J'^2-M'^2}{4J'^2-1}}\delta_{J,J'-1}.$$

Note that in the first term the quantum numbers are J,M, and in the second term they are J',M'.

9. If the dipole moment has a quadratic dependence on the separation of the harmonic oscillator, x, in a power series expansion of the dipole moment,

$\mu = \mu_o + ax + bx^2 + \ldots, b \neq 0.$

The transition from n = 0 to n = 2 is allowed only if $\langle\psi_o|\,\mu(x)\psi_2\rangle \neq 0$.

$$\langle\psi_o|\mu(x)\psi_2\rangle = \langle\psi_o|bx^2\,\psi_2\rangle$$
$$= \int_{-\infty}^{\infty}\left(\frac{\beta^2}{\pi}\right)^{\frac{1}{4}} e^{-\beta^2x^2/2} bx^2 \left(\frac{\beta^2}{64\pi}\right)^{\frac{1}{4}} (4\beta^2x^2 - 2)e^{-\beta^2x^2/2}dx$$
$$= \left(\frac{\beta^2}{\pi}\right)^{\frac{1}{2}} \frac{b}{4\sqrt{2}}\left[2\beta^2\int_{-\infty}^{\infty}x^4e^{-\beta^2x^2}dx - \int_{-\infty}^{\infty}x^2e^{-\beta^2x^2}dx\right]$$
$$= \left(\frac{\beta^2}{\pi}\right)^{\frac{1}{2}} \frac{b}{4\sqrt{2}}\left[2\beta^2\frac{3}{4}\sqrt{\frac{\pi}{\beta^{10}}} - \frac{1}{2}\sqrt{\frac{\pi}{\beta^6}}\right] = \frac{b}{2\sqrt{2}\beta^2}.$$

Thus $\langle\psi_o|\,\mu(x)\psi_2\rangle \neq 0$, and the transition is allowed if μ depends on x quadratically.

10. Applying the energy expression on p. 375, we obtain the energy of a 0,0 → n,0 transition:

$$\Delta E(n) = E_{n+1,0} - E_{0,0}$$
$$= \tilde{\omega}_e[(n+1/2)-(1/2)] - \tilde{\omega}_e\tilde{\chi}_e\left[(n+1/2)^2-(1/2)^2\right]$$
$$+ \tilde{\omega}_e\tilde{y}_e\left[[n+1/2)^3-(1/2)^3\right]$$
$$= \tilde{\omega}_e n - \tilde{\omega}_e\tilde{\chi}_e(n+n^2) + \tilde{\omega}_e\tilde{y}_e[(3/4)n+(3/2)n^2+n^3].$$

We can find the values of the vibrational frequency and the anharmonicity constants by evaluating this equation for each of the three given transitions and solving for the three unknowns.

$$\tilde{\omega}_e(1) - \tilde{\omega}_e\tilde{\chi}_e(2) + \tilde{\omega}_e\tilde{y}_e(13/4) = 1600.$$
$$\tilde{\omega}_e(2) - \tilde{\omega}_e\tilde{\chi}_e(6) + \tilde{\omega}_e\tilde{y}_e(31/2) = 3100.$$
$$\tilde{\omega}_e(3) - \tilde{\omega}_e\tilde{\chi}_e(12) + \tilde{\omega}_e\tilde{y}_e(171/4) = 4503.$$

$\tilde{\omega}_e = 1702.875\text{cm}^{-1}$, $\tilde{\omega}_e\tilde{\chi}_e = 52.250\text{cm}^{-1}$, $\tilde{\omega}_e\tilde{y}_e = 0.500\text{cm}^{-1}$.

Neglecting the second anharmonicity constant, and looking just at the first two transitions, we can find the vibrational frequency and the first anharmonicity constant by evaluating the equation (neglecting the second anharmonicity) and solving two equations in two unknowns:

$$\tilde{\omega}_e(1) - \tilde{\omega}_e\tilde{\chi}_e(2) = 1600.$$

$$\tilde{\omega}_e(2) - \tilde{\omega}_e\tilde{\chi}_e(6) = 3100.$$

$$\tilde{\omega}_e = 1700.0 \text{ cm}^{-1} \text{ and } \tilde{\omega}_e\tilde{\chi}_e = 50.0 \text{ cm}^{-1}.$$

11. We found the relationship between the rotational constant, reduced mass, and equilibrium bond length in Problem 2:

$$\tilde{B}_e = \frac{2.799 \times 10^{-46} \text{ cm}^{-1} \text{ kg m}^2}{\mu r_e^2}.$$

$$\mu_{CO} = (12.00)(15.99491) / (12.00 + 15.99491) = 6.8562 \text{ amu}.$$

$$r_e = \left(\frac{2.799 \times 10^{-46} \text{ cm}^{-1} \text{ kg m}^2}{(6.8562 \text{ amu})(1.661 \times 10^{-27} \text{ kg amu}^{-1})(1.9313 \text{ cm}^{-1})}\right)^{1/2}$$

$$= 1.128 \times 10^{-10} \text{ m} = 1.128 \text{ Å}.$$

It is assumed that the equilibrium bond length is unchanged in different isotopes, so now we use this bond length to calculate the reduced mass for $^{13}C^{17}O$.

$$\mu = (13.0034)(16.9991) / (13.0034 + 16.9991) = 7.3676 \text{ amu}.$$

$$\tilde{B}_e = \left(\frac{2.799 \times 10^{-46} \text{ cm}^{-1} \text{ kg m}^2}{(7.3676 \text{ amu})(1.661 \times 10^{-27} \text{ kg amu}^{-1})(1.128 \times 10^{-10} \text{ m})^2}\right)$$

$$= 1.7976 \text{ cm}^{-1}.$$

We can find the harmonic frequency of $^{13}C^{17}O$ from that of CO and the reduced masses of each:

$$\tilde{\omega}_e(^{13}C^{17}O) = \tilde{\omega}_e(CO)\sqrt{\frac{\mu(CO)}{\mu(^{13}C^{17}O)}} = 2170.2\sqrt{\frac{6.8562}{7.3676}} = 2093.5\ cm^{-1}.$$

Now we can write the transition energies for the R-branch of $^{13}C^{17}O$ as

$$\Delta E_{0,J\to 0,J+1} = 2093.5 + 2(J+1)(1.80)\ (cm^{-1})$$

and those of the P-branch as

$$\Delta E_{0,J+1\to 0,J} = 2093.5 - 2(J+1)(1.80)\ (cm^{-1})$$

Assignment	Line position (cm^{-1})
$J = 4\to 3$	2079.1
$J = 3\to 2$	2082.7
$J = 2\to 1$	2086.3
$J = 1\to 0$	2089.9
$J = 0\to 1$	2097.1
$J = 1\to 2$	2100.7
$J = 2\to 3$	2104.3
$J = 3\to 4$	2107.9

12. The values listed in Table 9.1 pertain to the most abundant isotope, which in the case of HCl is $H^{35}Cl$. Equations [9-35] and [9-39] relate the different spectroscopic constants to the physical properties. The key to relating one isotope to another is that the potential energy curve (i.e., the force constant) is unchanged; only the reduced mass differs.

$$\mu(H^{35}Cl) = (1.00785)(34.9689) / (1.00785 + 34.9689) = 0.9796\ \text{amu}.$$

$$\mu(H^{37}Cl) = (1.00785)(36.9659) / (1.00785 + 36.9659) = 0.9811\ \text{amu}.$$

$$\tilde{\omega}_e(H^{37}Cl) = \tilde{\omega}_e(H^{35}Cl)\sqrt{\frac{\mu(H^{35}Cl)}{\mu(H^{37}Cl)}} = 2989.7\sqrt{\frac{0.9796}{0.9811}} = 2987.4\ cm^{-1}.$$

Examination of [9-35] reveals the dependence of the rotational constant, centrifugal distortion constant, and the vibration-rotation coupling term in terms of the reduced mass. The rotational constant is inversely proportional to μ, so

$$\tilde{B}_e(H^{37}Cl) = \tilde{B}_e(H^{35}Cl)\frac{\mu(H^{35}Cl)}{\mu(H^{37}Cl)} = 10.59\frac{0.9796}{0.9811} = 10.57 \text{ cm}^{-1}.$$

The vibration-rotation coupling constant is inversely dependent on the reduced mass squared and the vibrational frequency:

$$\tilde{\alpha}(H^{37}Cl) = \tilde{\alpha}(H^{37}Cl)\left(\frac{\omega\mu^2(H^{35}Cl)}{\omega\mu^2(H^{37}Cl)}\right) = 0.3019\left(\frac{(2989.7)(0.9796)^2}{(2987.4)(0.9811)^2}\right)$$
$$= 0.3012 \text{ cm}^{-1}.$$

Equation [9-39] shows that the anharmonicity constant is inversely proportional to the reduced mass.

$$\tilde{\omega}\tilde{\chi}_e(H^{37}Cl) = \tilde{\omega}\tilde{\chi}_e(H^{35}Cl)\frac{\mu(H^{35}Cl)}{\mu(H^{37}Cl)} = 52.05\frac{0.9796}{0.9811} = 51.97 \text{ cm}^{-1}.$$

The energy expression for $H^{35}Cl$ is

$$E_{n,J}(\text{cm}^{-1}) = 2989.7\ (n + 1/2) - 52.05\ (n + 1/2)^2 + 10.59\ J\ (J + 1)$$
$$- 0.3019\ (n + 1/2)\ J\ (J + 1).$$

For an R–branch transition, the energy is given by

$$E_{1,J+1} - E_{0,J} =$$

$$2989.7(3/2) - 52.05(3/2)^2 + 10.59(J+1)(J+2) - 0.3019(3/2)(J+1)(J+2)$$
$$- [2989.7(1/2) - 52.05(1/2)^2 + 10.59J(J+1) - 0.3019(1/2)J(J+1)]$$
$$= 2885.6 + 21.18(J+1) - 0.3019(J+1)(J+3).$$

The P-branch transitions are given by

$$E_{1,J-1} - E_{0,J} = 2989.7(3/2) - 52.05(3/2)^2 + 10.59(J-1)J - 0.3019(3/2)(J-1)J$$
$$- [2989.7(1/2) - 52.05(1/2)^2 + 10.59J(J+1) - 0.3019(1/2)J(J+1)]$$
$$= 2885.6 - 21.18J - 0.3019J(J-2).$$

For $H^{37}Cl$:

$$E_{n,J}(cm^{-1}) = 2987.4\ (n + 1/2) - 51.97\ (n + 1/2)^2 + 10.57\ J\ (J + 1) - 0.3012\ (n + 1/2)\ J\ (J + 1).$$

For an R-branch transition, the energy is given by

$$E_{1,J+1} - E_{0,J} = 2987.4(3/2) - 51.97(3/2)^2 + 10.57(J+1)(J+2) - 0.3012(3/2)(J+1)(J+2) - [2987.4(1/2) - 51.97(1/2)^2 + 10.57J(J+1) - 0.3012(1/2)J(J+1)]$$

$$= 2883.5 + 21.14(J+1) - 0.3012(J+1)(J+3).$$

The P-branch transitions are given by

$$E_{1,J-1} - E_{0,J} = 2987.4(3/2) - 51.97(3/2)^2 + 10.57(J-1)J - 0.3012(3/2)(J-1)J - [2987.4(1/2) - 51.97(1/2)^2 + 10.57J(J+1) - 0.3012(1/2)J(J+1)]$$

$$= 2883.5 - 21.14J - 0.3012J(J-2).$$

Transition	$H^{35}Cl$ (cm^{-1})	$H^{37}Cl$ (cm^{-1})
$J = 4 \rightarrow 3$	2798.5	2796.5
$J = 3 \rightarrow 2$	2821.2	2819.1
$J = 2 \rightarrow 1$	2843.2	2841.2
$J = 1 \rightarrow 0$	2864.7	2862.6
$J = 0 \rightarrow 1$	2905.9	2903.7
$J = 1 \rightarrow 2$	2925.5	2923.4
$J = 2 \rightarrow 3$	2944.6	2942.4
$J = 3 \rightarrow 4$	2963.1	2960.8
$J = 4 \rightarrow 5$	2980.9	2978.7

13. $$E_{n+1,J+1} - E_{n,J} = \left(n + \frac{3}{2}\right)\hbar\omega + B(J+1)(J+2) - D(J+1)^2(J+2)^2 - \left[\left(n + \frac{1}{2}\right)\hbar\omega + B(J)(J+1) - DJ^2(J+1)^2\right]$$

$$= \hbar\omega + 2B(J+1) - 4D(J+1)^3.$$

$$E_{n+1,J-1} - E_{n,J} = (n + \frac{3}{2})\hbar\omega + B(J-1)(J) - D(J-1)^2(J)^2$$
$$-\left((n + \frac{1}{2})\hbar\omega + B(J)(J+1) - DJ^2(J+1)^2\right)$$
$$= \hbar\omega - 2BJ + 4DJ^3.$$

14. $$E_{1,1} - E_{0,0} = (3/2)\tilde{\omega} + \tilde{B}(1)(2) - \tilde{\alpha}(3/2)(1)(2) - \tilde{D}(1)^2(2)^2$$
$$-\left[(1/2)\tilde{\omega} + \tilde{B}(0)(1) - \tilde{\alpha}(1/2)(0)(1) - \tilde{D}(0)^2(1)^2\right]$$
$$= \tilde{\omega} + 2\tilde{B} - 3\tilde{\alpha} - 4\tilde{D}.$$

$$E_{1,0} - E_{0,1} = (3/2)\tilde{\omega} + \tilde{B}(0)(1) - \tilde{\alpha}(3/2)(0)(1) - \tilde{D}(0)^2(1)^2$$
$$-\left[(1/2)\tilde{\omega} + \tilde{B}(1)(2) - \tilde{\alpha}(1/2)(1)(2) - \tilde{D}(1)^2(2)^2\right]$$
$$= \tilde{\omega} - 2\tilde{B} + \tilde{\alpha} + 4\tilde{D}.$$

$$\Delta E(P \rightarrow R) = (E_{1,1} - E_{0,0}) - (E_{1,0} - E_{0,1}) = 4\tilde{B} - 4\tilde{\alpha} - 8\tilde{D}.$$

15. From the energy expression, the rotational constant can be identified as 7.513 cm^{-1} [the constant associated with J(J+1)]. From this value, as in Problem 11, we can be calculate the bond length:

$$\mu_{LiH} = (1.00785)(\ 7.01600) / (1.00785 + 7.01600) = 0.88126 \text{ amu}.$$

$$r_e = \left(\frac{2.799 \times 10^{-46}\ cm^{-1} kg\ m^2}{(0.88126\ amu)(1.661 \times 10^{-27}\ kg\ amu^{-1})(7.513\ cm^{-1})}\right)^{1/2}$$

$$= 1.595 \times 10^{-10}\ m = 1.595\ \text{Å}.$$

Now, to convert all the spectroscopic constants to those of LiD, we follow Problem 12.

$$\mu_{LiD} = (2.01410)(\ 7.01600) / (2.01410 + 7.01600) = 1.5648 \text{ amu}.$$

$$\tilde{\omega}_e(\text{LiD}) = \tilde{\omega}_e(\text{LiH})\sqrt{\frac{\mu(\text{LiH})}{\mu(\text{LiD})}} = 1405.65\sqrt{\frac{0.88126}{1.5648}} = 1054.9\ \text{cm}^{-1}.$$

$$\tilde{B}_e(\text{LiD}) = \tilde{B}_e(\text{LiH})\frac{\mu(\text{LiH})}{\mu(\text{LiD})} = 7.513\frac{0.88126}{1.5648} = 4.231\ \text{cm}^{-1}.$$

$$\tilde{\alpha}(\text{LiD}) = \tilde{\alpha}(\text{LiH})\left(\frac{\omega\mu^2(\text{LiH})}{\omega\mu^2(\text{LiD})}\right) = 0.213\left(\frac{(1405.65)(0.88126)^2}{(1054.9)(1.5648)^2}\right) = 0.090\ \text{cm}^{-1}.$$

$$\tilde{\omega}\tilde{\chi}_e(\text{LiD}) = \tilde{\omega}\tilde{\chi}_e(\text{LiH})\frac{\mu(\text{LiH})}{\mu(\text{LiD})} = 23.20\frac{0.88126}{1.5648} = 13.07\ \text{cm}^{-1}.$$

The centrifugal constant is inversely dependent on the square of the reduced mass [9-35]:

$$\tilde{D}(\text{LiD}) = \tilde{D}(\text{LiH})\left(\frac{\mu(\text{LiH})}{\mu(\text{LiD})}\right)^2 = 0.01\left(\frac{0.88126}{1.5648}\right)^2 = 0.003\ \text{cm}^{-1}.$$

The energy expression for LiD is now

$$E_{n,J}(\text{cm}^{-1}) = 1054.9\,(n + 1/2) - 13.07\,(n + 1/2)^2 + 4.321\,J\,(J + 1)$$

$$- 0.090\,(n + 1/2)\,J\,(J + 1) - 0.003\,[J\,(J + 1)]^2.$$

16. Using the energy expressions from Problem 13, we can generate the table of line positions and predictions based on the vibrational frequency, rotational constant, and centrifugal distortion term.

Assignments	Line position (cm^{-1})	Predicted value
$J = 4 \to 3$	2028	$\omega - 8B + 256D$
$J = 3 \to 2$	2045	$\omega - 6B + 108D$
$J = 2 \to 1$	2063	$\omega - 4B + 32D$
$J = 1 \to 0$	2082	$\omega - 2B + 4D$
$J = 0 \to 1$	2122	$\omega + 2B - 4D$
$J = 1 \to 2$	2141	$\omega + 4B - 32D$
$J = 2 \to 3$	2159	$\omega + 6B - 108D$
$J = 3 \to 4$	2176	$\omega + 8B - 256D$

ω, B, and D could be evaluated by choosing three particular data points, or by least squares fitting. The least squares fitting approach determines $\omega = 2102$, $B = 9.413$ and $D = 0.021$; the root-mean-square error from these values is only 0.17 cm^{-1}. The fit is shown in the figure below (the line represents the values of the fitted function, the solid dots the actual values).

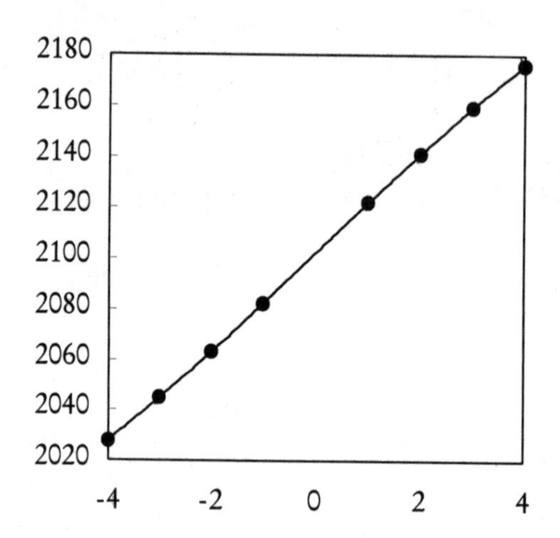

17. The qualitative forms of the vibrational normal modes will be similar; however, the infrared activity will be different. For CO_2, the symmetric stretch mode is infrared inactive; the dipole moment does not change (it remains identically zero by symmetry). For NNO, all modes will be infrared active, as every mode leads to a change in the dipole moment.

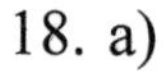

18. a)

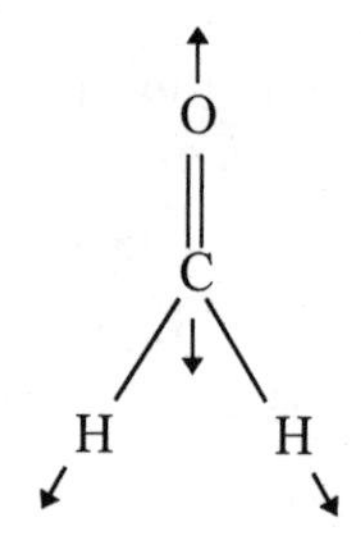

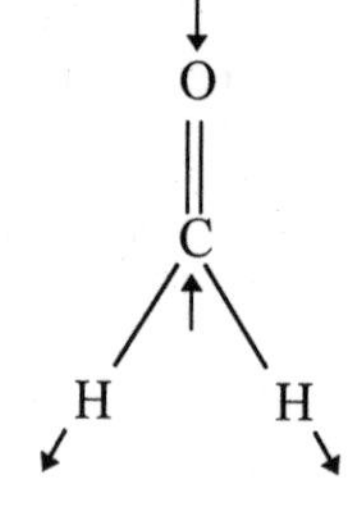

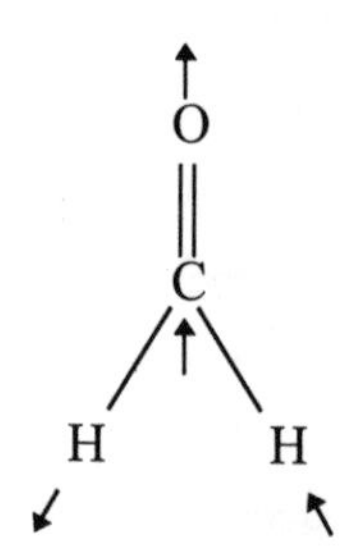

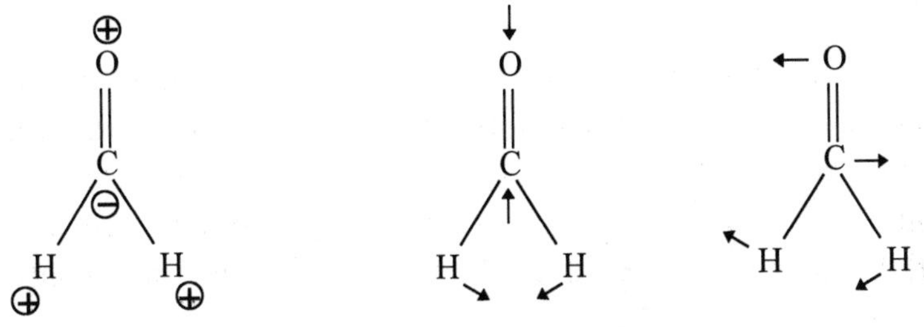

b)

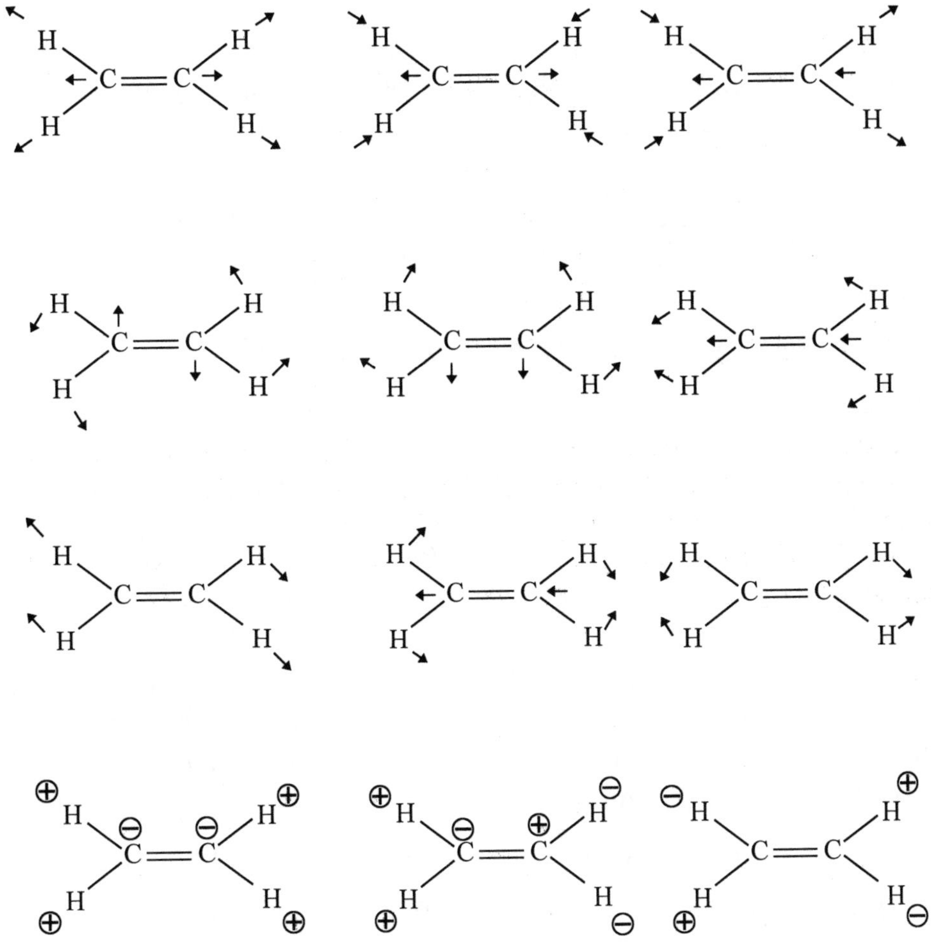

c)

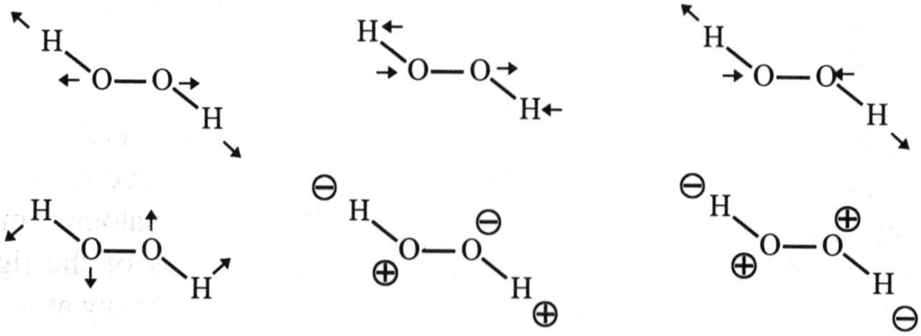

[Note: The circles represent motion into and out of the plane.]

In constructing these motions, the two constraints which must be met are that each motion be orthogonal to all others (i.e., that they are completely different from each other) and that the center of mass remain fixed.

19. The zero-point energies (Z.P.E.) within the harmonic approximation are merely the sum of 1/2 $\hbar\omega$ for each vibrational mode. The harmonic frequencies for all three molecules are given in Table 9.2.

Z.P.E.(water) = $1/2\ (3657 + 3756 + 1595) = 4504\ \text{cm}^{-1}$.

Z.P.E.(acetylene) = $(1/2)(1974 + 3289 + 3374 + (2 \times 612) + (2 \times 730))$
$= 5661\ \text{cm}^{-1}$
(for linear molecules, the bending modes are doubly degenerate).

Z.P.E.(formaldehyde) = $1/2\ (1746 + 2783 + 2843 + 1500 + 1167 + 1249)$
$= 5644\ \text{cm}^{-1}$.

20. The frequency for this "pseudodiatomic" depends on the appropriate reduced mass. For H–CN, the appropriate masses to consider are the hydrogen atom and the combined mass of the carbon and nitrogen atom.

$\mu(\text{HCN}) = (1.0078)(12.00+14.003)/(1.0078+12.00+14.003) = 0.9702$.

$\mu(\text{H}^{13}\text{CN}) = (1.0078)(13.003+14.003)/(1.0078+13.003+14.003) = 0.9715$.

$\omega(\text{H}^{13}\text{CN}) = 3000(0.9702/0.9715)^{1/2} = 2998\ \text{cm}^{-1}$.

$\mu(\text{HCCCN}) = (1.0078)(3\times12.00+14.003)/(1.0078+3\times12.00+14.003)$
$=0.9879$.

$\omega(\text{HCCCN}) = 3000(0.9702/0.9879)^{1/2} = 2973\ \text{cm}^{-1}$.

$\mu(DCN) = (2.014)(12.00+14.003)/(2.014+12.00+14.003) = 1.8692.$

$\omega(DCN) = 3000(0.9702/1.8692)^{1/2} = 2161\ cm^{-1}.$

From Table 9.2 we can see that the H–C frequency of HCN is 3311 cm^{-1}, and that of DCN is 2630. The reduction in the frequency due to isotopic substitution is less than that predicted by the pseudodiatomic model presented above, but it illustrates that increasing the mass of the lighter mass is far more significant than increasing the mass of the heavy atom.

21. A sketch of the low-lying vibrational energy levels of HCCH and DCCD is shown below. All the energy levels (in cm^{-1}) for HCCH less than 2500 cm^{-1} are plotted, and the associated energy states (labeled by the quantum number

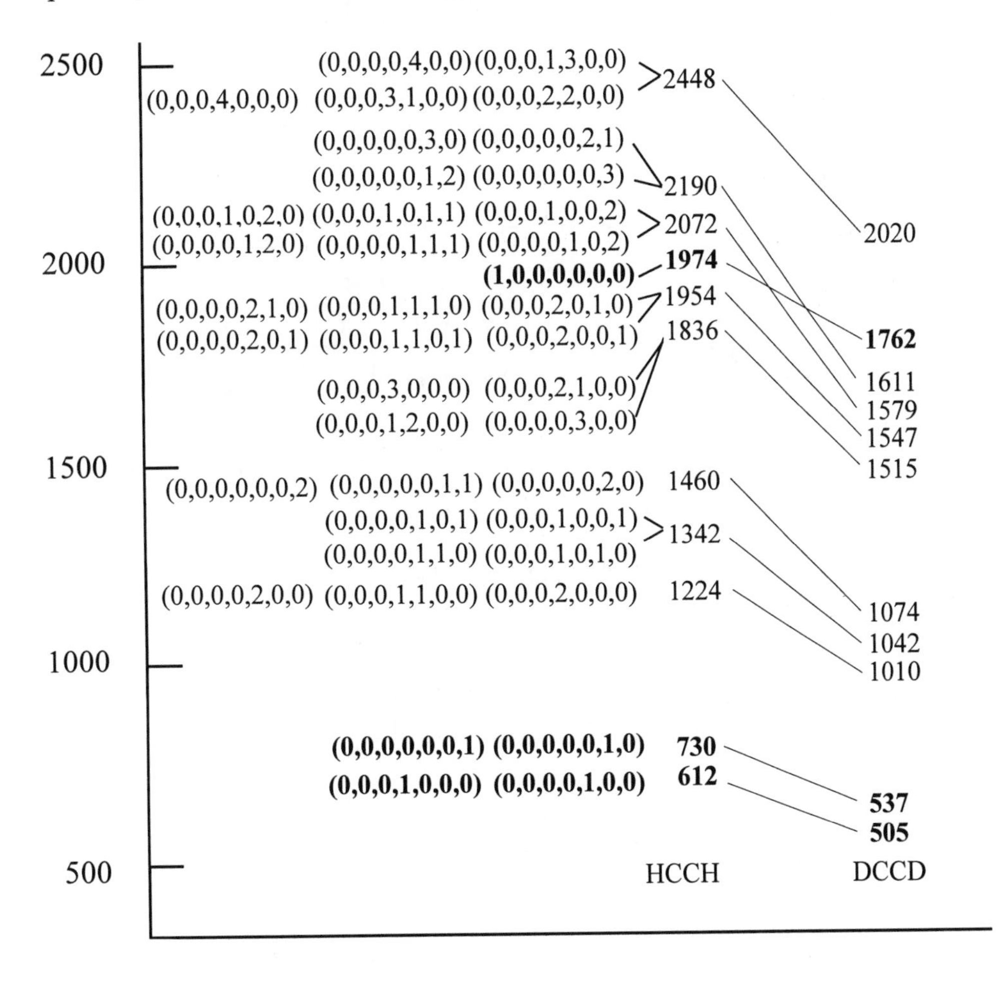

for each of the seven vibrational modes) are indicated. The quantum numbers follow the order of the vibrational frequencies as listed in Table 9.2 in the text. The fundamental transitions are marked in bold type.

22. This represents a perturbation to a harmonic oscillator, where k and m = 1. For these values, $\omega = 1$, and so the zero-order energies are simply $1/2\,\hbar$, $3/2\,\hbar$, $5/2\,\hbar$, $7/2\,\hbar$, and $9/2\,\hbar$ for n = 0, 1, 2, 3, and 4.

The first-order correction is just the expectation value of the perturbing Hamiltonian. Note that since m = 1 and k = 1, $\beta^2 = 1/\hbar$.

$$E_0^{(1)} = \left(\frac{\beta}{\sqrt{\pi}}\right)\int_{-\infty}^{\infty} e^{-\beta^2x^2/2}\,\frac{3}{2}e^{-\beta^2x^2/2}e^{-\beta^2x^2/2}dx$$

$$= \frac{3}{2}\left(\frac{\beta}{\sqrt{\pi}}\right)\int_{-\infty}^{\infty} e^{-3\beta^2x^2/2}dx = \frac{3}{2}\left(\frac{\beta}{\sqrt{\pi}}\right)\sqrt{\frac{2\pi}{3\beta^2}} = \sqrt{\frac{3}{2}}.$$

$$E_1^{(1)} = \frac{\beta}{2\sqrt{\pi}}\int_{-\infty}^{\infty} 2\beta x e^{-\beta^2x^2/2}\,\frac{3}{2}e^{-\beta^2x^2/2}\,2\beta x e^{-\beta^2x^2/2}dx$$

$$= \frac{3\beta^3}{\sqrt{\pi}}\int_{-\infty}^{\infty} x^2 e^{-3\beta^2x^2/2}dx = \frac{3\beta^3}{\sqrt{\pi}}\frac{1}{2}\sqrt{\frac{\pi}{(3\beta^2/2)^3}} = \sqrt{\frac{2}{3}}.$$

$$E_2^{(1)} = \frac{\beta}{8\sqrt{\pi}}\int_{-\infty}^{\infty}(4\beta^2x^2-2)e^{-\beta^2x^2/2}\,\frac{3}{2}e^{-\beta^2x^2/2}(4\beta^2x^2-2)e^{-\beta^2x^2/2}dx$$

$$= \frac{3\beta}{16\sqrt{\pi}}\int_{-\infty}^{\infty}\left(16\beta^4x^4-16\beta^2x^2+4\right)e^{-3\beta^2x^2/2}dx$$

$$= \frac{3\beta}{16\sqrt{\pi}}\left[16\beta^4\,\frac{3}{4}\left(\frac{2}{3\beta^2}\right)^{5/2} - \frac{16\beta^2}{2}\left(\frac{2}{3\beta^2}\right)^{3/2} + 4\left(\frac{2}{3\beta^2}\right)^{1/2}\right]\sqrt{\pi}$$

$$= \sqrt{\frac{2}{3}} - \sqrt{\frac{2}{3}} + \frac{3}{4}\sqrt{\frac{2}{3}} = \frac{3}{4}\sqrt{\frac{2}{3}}.$$

$$E_3^{(1)} = \frac{\beta}{48\sqrt{\pi}} \int_{-\infty}^{\infty} (8\beta^3 x^3 - 12\beta x) e^{-\beta^2 x^2/2} \frac{3}{2} e^{-\beta^2 x^2/2} (8\beta^3 x^3 - 12\beta x) e^{-\beta^2 x^2/2} dx$$

$$= \frac{\beta}{32\sqrt{\pi}} \int_{-\infty}^{\infty} \left(64\beta^6 x^6 - 192\beta^4 x^4 + 144\beta^2 x^2\right) e^{-3\beta^2 x^2/2} dx =$$

$$= \frac{\beta}{\sqrt{\pi}} \left[2\beta^6 \frac{15}{8} \left(\frac{2}{3\beta^2}\right)^{7/2} - 6\beta^4 \frac{3}{4} \left(\frac{2}{3\beta^2}\right)^{5/2} + \frac{(9/2)\beta^2}{2} \left(\frac{2}{3\beta^2}\right)^{3/2} \right] \sqrt{\pi}$$

$$= \frac{10}{9}\sqrt{\frac{2}{3}} - 2\sqrt{\frac{2}{3}} + \frac{3}{2}\sqrt{\frac{2}{3}} = \frac{11}{18}\sqrt{\frac{2}{3}}.$$

$$E_4^{(1)} = \frac{\beta}{384\sqrt{\pi}} \int_{-\infty}^{\infty} \left[(16\beta^4 x^4 - 48\beta^2 x^2 + 12) e^{-\beta^2 x^2/2} \right]^2 \frac{3}{2} e^{-\beta^2 x^2/2} dx$$

$$= \frac{\beta}{\sqrt{\pi}} \int_{-\infty}^{\infty} \left[\beta^8 x^8 - 6\beta^6 x^6 + (21/2)\beta^4 x^4 - (9/2)\beta^2 x^2 + (9/16) \right] e^{-3\beta^2 x^2/2} dx$$

$$= \frac{\beta}{\sqrt{\pi}} \left[\beta^8 \frac{105}{16} \left(\frac{2}{3\beta^2}\right)^{9/2} - 6\beta^6 \frac{15}{8} \left(\frac{2}{3\beta^2}\right)^{7/2} + (21/2)\beta^4 \frac{3}{4} \left(\frac{2}{3\beta^2}\right)^{5/2} \right.$$

$$\left. - (9/2)\beta^2 \frac{1}{2} \left(\frac{2}{3\beta^2}\right)^{3/2} + (9/16) \left(\frac{2}{3\beta^2}\right)^{1/2} \right] \sqrt{\pi}$$

$$= \frac{35}{27}\sqrt{\frac{2}{3}} - \frac{10}{3}\sqrt{\frac{2}{3}} + \frac{7}{2}\sqrt{\frac{2}{3}} - \frac{3}{2}\sqrt{\frac{2}{3}} + \frac{9}{16}\sqrt{\frac{2}{3}} = \frac{227}{432}\sqrt{\frac{2}{3}}.$$

To calculate the second-order energy corrections, we have to evaluate all the nonzero terms of the form $\langle \Psi_j | H^{(1)} \Psi_i \rangle$; here, $H^{(1)}$ is the gaussian perturbation. Since the perturbation is an even function, the only nonzero elements are $\langle \Psi_0 | H^{(1)} \Psi_2 \rangle$, $\langle \Psi_0 | H^{(1)} \Psi_4 \rangle$, $\langle \Psi_2 | H^{(1)} \Psi_4 \rangle$, and $\langle \Psi_1 | H^{(1)} \Psi_3 \rangle$ (all other combinations result in integrals over odd functions, which are identically zero).

$$\left\langle \Psi_0 | H^{(1)} \Psi_2 \right\rangle = \int_{-\infty}^{\infty} \left(\frac{\beta}{\sqrt{\pi}}\right)^{1/2} e^{-\beta^2 x^2/2} \frac{3}{2} e^{-\beta^2 x^2/2} \left(\frac{\beta}{8\sqrt{\pi}}\right)^{1/2} (4\beta^2 x^2 - 2) e^{-\beta^2 x^2/2} dx$$

$$= \frac{3\beta}{4\sqrt{2\pi}} \int_{-\infty}^{\infty} (4\beta^2 x^2 - 2) e^{-3\beta^2 x^2/2} dx$$

$$= \frac{3\beta}{4\sqrt{2\pi}} \left[\frac{4\beta^2}{2} \left(\frac{2}{3\beta^2}\right)^{3/2} - 2\left(\frac{2}{3\beta^2}\right)^{1/2} \right] \sqrt{\pi}$$

$$= \sqrt{\frac{1}{3}} - \sqrt{\frac{2}{3}} = -\frac{1}{2\sqrt{3}}.$$

$$\left\langle \Psi_0 | H^{(1)} \Psi_4 \right\rangle$$

$$= \frac{\beta}{8\sqrt{6\pi}} \int_{-\infty}^{\infty} e^{-\beta^2 x^2/2} \frac{3}{2} e^{-\beta^2 x^2/2} (16\beta^4 x^4 - 48\beta^2 x^2 + 12) e^{-\beta^2 x^2/2} dx$$

$$= \frac{3\beta}{16\sqrt{6\pi}} \int_{-\infty}^{\infty} (16\beta^4 x^4 - 48\beta^2 x^2 + 12) e^{-3\beta^2 x^2/2} dx$$

$$= \frac{3\beta}{16\sqrt{6\pi}} \left[16\beta^4 \frac{3}{4} \left(\frac{2}{3\beta^2}\right)^{5/2} - 48\beta^2 \frac{1}{2} \left(\frac{2}{3\beta^2}\right)^{3/2} + 12\left(\frac{2}{3\beta^2}\right)^{1/2} \right] \sqrt{\pi}$$

$$= \frac{1}{3} - 1 + \frac{3}{4} = \frac{1}{12}.$$

$$\left\langle \Psi_2 | H^{(1)} \Psi_4 \right\rangle$$

$$= \frac{\beta}{32\sqrt{3\pi}} \int_{-\infty}^{\infty} (4\beta^2 x^2 - 2) e^{-\beta^2 x^2/2} \frac{3}{2} e^{-\beta^2 x^2/2} (16\beta^4 x^4 - 48\beta^2 x^2 + 12) e^{-\beta^2 x^2/2} dx$$

$$= \frac{3\beta}{64\sqrt{3\pi}} \int_{-\infty}^{\infty} (64\beta^6 x^6 - 224\beta^4 x^4 + 144\beta^2 x^2 - 24) e^{-3\beta^2 x^2/2} dx$$

$$= \frac{3\beta}{64\sqrt{3\pi}}\left[64\beta^6\frac{15}{8}\left(\frac{2}{3\beta^2}\right)^{7/2} - 224\beta^4\frac{3}{4}\left(\frac{2}{3\beta^2}\right)^{5/2} + 144\beta^2\frac{1}{2}\left(\frac{2}{3\beta^2}\right)^{3/2}\right.$$

$$\left. -24\left(\frac{2}{3\beta^2}\right)^{1/2}\right]\sqrt{\pi} = \frac{5}{9}\sqrt{2} - \frac{7}{6}\sqrt{2} + \frac{3}{4}\sqrt{2} - \frac{3}{8}\sqrt{2} = -\frac{17}{72}\sqrt{2}.$$

$$\left\langle\Psi_1|H^{(1)}\Psi_3\right\rangle = \frac{\beta}{4\sqrt{6\pi}}\int_{-\infty}^{\infty} 2\beta x e^{-\beta^2x^2/2}\,\frac{3}{2}e^{-\beta^2x^2/2}(8\beta^3x^3 - 12\beta x)e^{-\beta^2x^2/2}dx$$

$$= \frac{3\beta^3}{4\sqrt{6\pi}}\int_{-\infty}^{\infty}\left(8\beta^2x^4 - 12x^2\right)e^{-3\beta^2x^2/2}dx$$

$$= \frac{3\beta^3}{4\sqrt{6\pi}}\left[8\beta^2\frac{3}{4}\left(\frac{2}{3\beta^2}\right)^{5/2} - 12\frac{1}{2}\left(\frac{2}{3\beta^2}\right)^{3/2}\right]\sqrt{\pi} = \frac{2}{3} - 1 = -\frac{1}{3}.$$

Now we can evaluate the second-order energy corrections.

$$E_0^{(2)} = \sum_j \frac{\left\langle\Psi_0|H^{(1)}\Psi_j\right\rangle^2}{E_0^{(0)} - E_j^{(0)}} = \frac{\left(-1/2\sqrt{3}\right)^2}{[(1/2)-(5/2)]\hbar} + \frac{(1/12)^2}{[(1/2)-(9/2)]\hbar} = -\frac{25}{576\hbar}.$$

$$E_1^{(2)} = \frac{(-1/3)^2}{[(3/2)-(7/2)]\hbar} = -\frac{1}{18\hbar}.$$

$$E_2^{(2)} = \frac{\left(-1/2\sqrt{3}\right)^2}{[(5/2)-(1/2)]\hbar} + \frac{\left(-17\sqrt{2}/72\right)^2}{[(5/2)-(9/2)]\hbar} = -\frac{73}{5184\hbar}.$$

$$E_3^{(2)} = \frac{(-1/3)^2}{[(7/2)-(3/2)]\hbar} = \frac{1}{18\hbar}.$$

$$E_4^{(2)} = \frac{(1/12)^2}{((9/2)-(1/2))\hbar} + \frac{\left(-17\sqrt{2}/72\right)^2}{((9/2)-(5/2))\hbar} = \frac{149}{2592\hbar}.$$

Chapter 10

Electronic Structure

Exercises

1. The energy of a one–electron atom is given by [10-9]

$$E_{nl} = -\frac{\mu Z^2 e^4}{2\hbar^2 n^2}.$$

For the n = 1 energy state of a hydrogen atom, where the proton mass is taken to be infinite, the reduced mass is just the mass of the electron. To see that this is true, we write the formula for the reduced mass.

$$\mu = \frac{m_e m_p}{m_e + m_p} = \frac{m_e \times \infty}{m_e + \infty} = \frac{m_e \times \infty}{\infty} = m_e.$$

To calculate the energy, we substitute in Z = 1 (the charge of the proton), n = 1, $\mu = m_e$, and the other physical constants.

$$E_{nl} = -\frac{(9.10939 \times 10^{-31}\ \text{kg})(1)^2(1.5189 \times ^{-14}\ \text{m}^{3/2}\text{kg}^{1/2}\text{s}^{-1})^4}{(2)(1.054537 \times 10^{-34}\ \text{J s})^2(1)^2}$$

$$= -2.1799 \times 10^{-18}\text{ J}.$$

Note the units for the charge on the electron; coulombs are not the correct units. If 1.609×10^{-19} C is used, it will lead to a final answer with units of kg C^4 $(Js)^{-2}$, not J. The value used in the equation was derived from the value given inside the front cover of the text of e = 4.803×10^{-10} $cm^{3/2}$ $g^{1/2}$ s^{-1}. These are units of charge called electrostatic units (*esu*). Checking the units to be correct is a useful and easy check on your work.

Using the conversion factors in Table VI.3, we convert this energy to cm^{-1}:

$$-2.1799 \times 10^{-18}\text{ J} \times 5.0341 \times 10^{22}\text{ cm}^{-1}\text{ J}^{-1} = -109{,}737\text{ cm}^{-1}.$$

This calculation is even simpler when *atomic units* are used [see p. 632 of Appendix VI]; in this system of units the numerical values of e, m_e, and $\hbar$ are all 1. The energy formula reduces (for the case of Z = 1 and n = 1) to simply –1/2. The atomic unit of energy is the *hartree*. Conversion factors relating it to J, cm^{-1}, and kcal mol^{-1} are given in Appendix VI.

2. $$\mu = \frac{m_e m_p}{m_e + m_p} = \frac{(9.1094 \times 10^{-31}\text{ kg})(1.6726 \times 10^{-27}\text{ kg})}{(9.1094 \times 10^{-31}\text{ kg} + 1.6726 \times 10^{-27}\text{ kg})} = 9.1044 \times 10^{-31}\text{ kg}.$$

Now we can use the result from Problem 1:

$$E_{100,0} = \left(\frac{9.1044 \times 10^{-31}\text{ kg}}{9.1094 \times 10^{-31}\text{ kg}}\right)\frac{(-109{,}737\text{ cm}^{-1})}{100^2} = -10.9677\text{ cm}^{-1}.$$

We can calculate the expectation value of r with [10-10]:

$$\langle r \rangle_{100,0} = (0.5295 \times 10^{-10}\text{ m})\frac{100^2}{1}\left(\frac{3}{2} + \frac{0(0+1)}{(2)(100)^2}\right) = 7.942 \times 10^{-7}\text{ m}.$$

Note: Here we use a = 0.529466×10^{-10} m, which applies to the actual hydrogen atom.

3. If we neglect the effect of the changes of the mass of the different nuclei, and simply use a_0 in [10-10], the only changes are due to the different charges of the nuclei.

He: $\langle r \rangle_{1,0} = 1.5 \times 0.529177 \times 10^{-10}$ m / 2 = 3.9688×10^{-11} m.

Ne: $\langle r \rangle_{1,0} = 1.5 \times 0.529177 \times 10^{-10}$ m / 10 = 7.9376×10^{-12} m.

Ar: $\langle r \rangle_{1,0} = 1.5 \times 0.529177 \times 10^{-10}$ m / 18 = 4.4098×10^{-12} m.

Kr: $\langle r \rangle_{1,0} = 1.5 \times 0.529177 \times 10^{-10}$ m / 36 = 2.2049×10^{-12} m.

To examine the effects of the changing reduced masses, we can calculate the actual value of a as defined by [10-11], and then recalculate $\langle r \rangle$.

^{4}He: mass = 4.0026 amu × 1.6605×10^{-27} kg amu^{-1} = 6.6463×10^{-27} kg.

$$\mu = \frac{(9.1094 \times 10^{-31}\ \text{kg})(6.6463 \times 10^{-27}\ \text{kg})}{(9.1094 \times 10^{-31}\ \text{kg} + 6.6463 \times 10^{-27}\ \text{kg})} = 9.1082 \times 10^{-31}\ \text{kg}.$$

$$a = \frac{m_e}{\mu} a_0 = \frac{9.1094 \times 10^{-27}\ \text{kg}}{9.1082 \times 10^{-27}\ \text{kg}} \times 0.529177 \times 10^{-10}\ \text{m}$$

$$= 0.52925 \times 10^{-10}\ \text{m}.$$

$\langle r \rangle_{1,0} = 1.5 \times a / Z = 1.5 \times 0.52925 \times 10^{-10}$ m / 2 = 3.9694×10^{-11} m.

^{20}Ne: mass = 19.9924 amu × 1.6605×10^{-27} kg amu^{-1} = 3.3197×10^{-26} kg.

$$\mu = \frac{(9.1094 \times 10^{-31}\ \text{kg})(3.3197 \times 10^{-26}\ \text{kg})}{(9.1094 \times 10^{-31}\ \text{kg} + 3.3197 \times 10^{-26}\ \text{kg})} = 9.1092 \times 10^{-31}\ \text{kg}.$$

$$a = \frac{m_e}{\mu} a_0 = \frac{9.1094 \times 10^{-27}\ \text{kg}}{9.1092 \times 10^{-27}\ \text{kg}} \times 0.529177 \times 10^{-10}\ \text{m}$$

$$= 0.52919 \times 10^{-10}\ \text{m}.$$

$\langle r \rangle_{1,0} = 1.5 \times a / Z = 1.5 \times 0.52919 \times 10^{-10}$ m / 10 = 7.9379×10^{-12} m.

^{36}Ar: mass = 35.9675 amu × 1.6605 × 10^{-27} kg amu^{-1} = 5.9724 × 10^{-26} kg.

$$\mu = \frac{(9.1094 \times 10^{-31}\ \text{kg})(5.9724 \times 10^{-26}\ \text{kg})}{(9.1094 \times 10^{-31}\ \text{kg} + 5.9724 \times 10^{-26}\ \text{kg})} = 9.1093 \times 10^{-31}\ \text{kg}.$$

$$a = \frac{m_e}{\mu} a_0 = \frac{9.1094 \times 10^{-27}\ \text{kg}}{9.1093 \times 10^{-27}\ \text{kg}} \times 0.529177 \times 10^{-10}\ \text{m}$$

$$= 0.52918 \times 10^{-10}\ \text{m}.$$

$$\langle r \rangle_{1,0} = 1.5 \times a / Z = 1.5 \times 0.52918 \times 10^{-10}\ \text{m} / 10 = 4.4099 \times 10^{-12}\ \text{m}.$$

^{84}Kr: mass = 83.9115 amu × 1.6605 × 10^{-27} kg amu^{-1} = 1.3934 × 10^{-25} kg.

$$\mu = \frac{(9.1094 \times 10^{-31}\ \text{kg})(1.3934 \times 10^{-25}\ \text{kg})}{(9.1094 \times 10^{-31}\ \text{kg} + 1.3934 \times 10^{-25}\ \text{kg})} = 9.1093 \times 10^{-31}\ \text{kg}.$$

$$a = \frac{m_e}{\mu} a_0 = \frac{9.1094 \times 10^{-27}\ \text{kg}}{9.1093 \times 10^{-27}\ \text{kg}} \times 0.529177 \times 10^{-10}\ \text{m}$$

$$= 0.52918 \times 10^{-10}\ \text{m}.$$

$$\langle r \rangle_{1,0} = 1.5 \times a / Z = 1.5 \times 0.52918 \times 10^{-10}\ \text{m} / 36 = 2.2049 \times 10^{-12}\ \text{m}.$$

The effect of the changes in the mass of the nucleus are much smaller than those of the changes in the charge.

4. For this occupancy, where the two electrons in the unfilled n = 2 shell are inequivalent (since they are in different spatial orbitals), the term symbols can be found by simply coupling the angular momentum of each. Let us designate the electron in the 2s orbital as 1, and the electron in the 2p orbital as 2. Then $l_1 = 0$ and $l_2 = 1$. The coupling rules from Chapter 9 showed that the possible values of $L_{tot} = l_1 + l_2, \ldots, |l_1 - l_2|$. In this case, the only possible value is $L_{tot} = 1$. The spin angular momentum of each electron is 1/2, so the possible values of S_{tot} are 1 and 0.

The total angular momentum, J, is found by coupling the orbital and spin angular momentum. Its values can be J = 2, 1, 0 (= 1 + 1, …, 1 – 1) and J = 1 (= 1 + 0). Thus, the term symbols for the states represented by this electron occupancy are $^3P_2$, $^3P_1$, $^3P_0$, and $^1P_1$.

5. If the plus and minus signs are interchanged, the spin function becomes $\alpha(1)\beta(2) + \alpha(2)\beta(1)$. To identify the spin state that this function represents, consider the possibilities whereby two particles with S = 1/2 could be coupled together; they can form S_{tot} = 1 or 0. The initial spin function, $\alpha(1)\beta(2) - \alpha(2)\beta(1)$, was identified with the singlet state, $S_{tot} = 0$. Because both of these functions have $m_s = 0$ and are orthogonal to one another, this new spin function must arise from the $S_{tot} = 1$ state, the triplet spin state.

Additional Exercises

6. First, we use [10-5] to find $L_3(z)$:

$$L_3 = e^z \frac{d^3}{dz^3}(z^3 e^{-z})$$

$$= e^z (6 - 18z + 9z^2 - z^3)\, e^{-z} = (6 - 18z + 9z^2 - z^3).$$

$$L_3^2 = \frac{d^2}{dz^2}(6 - 18z + 9z^2 - z^3) = 18 - 6z.$$

7. $$\langle R_{30} | R_{30} \rangle = \frac{N^2}{243} \int_0^\infty \left[(6 - 6\rho + \rho^2)e^{-\rho/2}\right]^2 r^2 dr$$

From Table 10.1, $\rho = 2Z\mu e^2 r / n\hbar^2$. For conciseness, we let $\rho = cr$. Then

$$\langle R_{30} | R_{30} \rangle = \frac{N^2}{243c^3} \int_0^\infty \left[(6 - 6\rho + \rho^2)e^{-\rho/2}\right]^2 \rho^2 d\rho$$

$$= \frac{N^2}{243c^3} \int_0^\infty \left(36\rho^2 - 72\rho^3 + 48\rho^4 - 12\rho^5 + \rho^6\right)e^{-\rho}\, d\rho$$

$$= \frac{N^2}{243c^3}[36 \times 2! - 72 \times 3! + 48 \times 4! - 12 \times 5! + 6!]$$

$$= \frac{N^2}{243c^3}[72 - 432 + 1152 - 1440 + 720] = \frac{8N^2}{27c^3}$$

$$= \left(\frac{8}{27}\right)\left(\frac{Z\mu e^2}{\hbar^2}\right)^3\left(\frac{3\hbar^2}{2Z\mu e^2}\right)^3 = 1.$$

8. As in the previous problem, let $\rho = cr$, where $c = 2Z\mu e^2 / n\hbar^2$.

$$R_{41}(r) = \frac{N}{32\sqrt{15}}\left(20cr - 10(cr)^2 + (cr)^3\right)e^{-cr/2}.$$

$$\frac{\partial R_{41}}{\partial r} = \frac{N}{32\sqrt{15}}\left[-\frac{c}{2}\left(20cr - 10(cr)^2 + (cr)^3\right) + \left(20c - 20c^2r + 3c^3r^2\right)\right]e^{-cr/2}$$

$$= \frac{N}{32\sqrt{15}}\left[20c - 30c^2r + 8c^3r^2 - c^4r^3/2\right]e^{-cr/2}.$$

$$\frac{\partial^2 R_{41}}{\partial r^2} = \frac{\partial}{\partial r}\frac{N}{32\sqrt{15}}\left[20c - 30c^2r + 8c^3r^2 - c^4r^3/2\right]e^{-cr/2}$$

$$= \frac{N}{32\sqrt{15}}\left[-\frac{c}{2}\left[20c - 30c^2r + 8c^3r^2 - c^4r^3/2\right] + \left(-30c^2 + 16c^3r - 3c^4r^2/2\right)\right]e^{-cr/2}$$

$$= \frac{N}{32\sqrt{15}}\left[-40c^2 + 31c^3r - \frac{11}{2}c^4r^2 + \frac{c^5r^3}{4}\right]e^{-cr/2}.$$

We can now use these results to simplify

$$-\frac{\hbar^2}{2\mu}\left(\frac{2}{r}\frac{\partial R}{\partial r} + \frac{\partial^2 R}{\partial r^2} - \frac{(1)(2)}{r^2}R\right)$$

$$= -\frac{\hbar^2}{2\mu}\left[\frac{2}{r}\left(20c - 30c^2r + 8c^3r^2 - \frac{c^4r^2}{2}\right) + \left(-40c^2 + 31c^3r - \frac{11c^4r^2}{2}\right.\right.$$

$$\left.\left. + \frac{c^5r^3}{4}\right) + \frac{2}{r^2}\left(20cr - 10c^2r^2 + c^3r^3\right)\right]e^{-cr/2}$$

$$= -\frac{c^2\hbar^2}{2\mu}\left(-80 + 45cr - \frac{13c^2r^2}{2} + \frac{c^3r^3}{4}\right)e^{-cr/2}. \qquad \text{(i)}$$

$$\frac{Ze^2}{r}R_{41} = cZe^2\left(20 - 10cr + c^2r^2\right)e^{-cr/2}. \qquad \text{(ii)}$$

Since $c = \dfrac{2Z\mu e^2}{n\hbar}$, $\dfrac{c^2\hbar^2}{2\mu} = \dfrac{Z^2\mu e^4}{8\hbar^2}$ and $cZe^2 = \dfrac{Z^2\mu e^4}{2\hbar^2}$. If we substitute these expressions into (i) and (ii), the left-hand side of [10-2] becomes

$$-\frac{\hbar^2}{2\mu}\left(\frac{2}{r}\frac{\partial R}{\partial r} + \frac{\partial^2 R}{\partial r^2} - \frac{(1)(2)}{r^2}R\right) - \frac{Ze^2R}{r}$$

$$= \left(\frac{Z^2\mu e^4}{8\hbar^2}\right)\left(5cr - \frac{5}{2}(cr)^2 + (cr)^3\right)e^{-cr/2}.$$

After multiplying the polynomial terms by 4 and dividing the constant by 4, we get

$$-\frac{\hbar^2}{2\mu}\left(\frac{2}{r}\frac{\partial R}{\partial r} + \frac{\partial^2 R}{\partial r^2} - \frac{(1)(2)}{r^2}R\right) - \frac{Ze^2R}{r} = \left(\frac{Z^2\mu e^4}{2(4)^4\hbar^2}\right)R_{41}.$$

This is exactly the energy predicted by [10-9], when $n = 4$.

9. The ionization energy for the hydrogen atom is simply the energy difference between the $n = \infty$ and the $n = 1$ state.

$$E_{\text{ionization}} = -\frac{R}{\infty^2} - \left(-\frac{R}{1}\right) = R,$$

where R is the Rydberg constant [see p. 9 in the text]. As we see from [10-9], the Rydberg constant depends on both the reduced mass and the charge of the nucleus. However, because the change in the reduced mass is small (see Problem 3), the Rydberg constant, and thus the ionization energy, for atoms other than hydrogen can simply be found as

$$E_{ionization}(X) = [Z(X)/Z(H)]^2 \times R_{hydrogen} = Z(X)^2 \times 109{,}677 \text{ cm}^{-1}.$$

Ion	$E_{ionization}(cm^{-1})$
He^+	438,960
Li^{2+}	987,660
C^{5+}	3,950,640
Ne^{9+}	10,974,000

All these transitions fall in the ultraviolet region of the spectrum.

10. The spin–orbit interaction is given by

$$E_{JLS}^{\text{spin-orbit}} = \frac{\gamma\hbar^2}{2}\left[J(J+1) - L(L+1) - S(S+1)\right].$$

$$E(^2P_{3/2}) = (\gamma\hbar^2/2)[(3/2)(5/2) - (1)(2) - (1/2)(3/2)] = \gamma\hbar^2/2.$$

$$E(^2P_{1/2}) = (\gamma\hbar^2/2)[(1/2)(3/2) - (1)(2) - (1/2)(3/2)] = -\gamma\hbar^2.$$

$$E(^2D_{3/2}) = (\gamma\hbar^2/2)[(3/2)(5/2) - (2)(3) - (1/2)(3/2)] = -3\gamma\hbar^2/2.$$

$$E(^2F_{5/2}) = (\gamma\hbar^2/2)[(5/2)(7/2) - (3)(4) - (1/2)(3/2)] = -2\gamma\hbar^2.$$

$$E(^2F_{7/2}) = (\gamma\hbar^2/2)[(7/2)(9/2) - (3)(4) - (1/2)(3/2)] = 3\gamma\hbar^2/2.$$

11. For an s^2 occupancy:

s_0	M_L	M_S		
↑↓	0	0	⇒	L = 0, S = 0

For a p^6 occupancy:

p_{-1}	p_0	p_1	M_L	M_S		
↑↓	↑↓	↑↓	0	0	⇒	L = 0, S = 0

For a d^{10} occupancy:

d_2	d_1	d_0	d_{-1}	d_{-2}	M_L	M_S		
↑↓	↑↓	↑↓	↑↓	↑↓	0	0	⇒	L = 0, S = 0

In each case, there is only one way to arrange the electrons, and so there is only one state of coupled angular momentum.

For a $1s^22p^1$ occupancy:

s_0	p_1	p_0	p_{-1}	M_L	M_S		
↑↓	↑			1	1/2	⇒	L = 1, S = 1/2
↑↓	↓			1	−1/2		
↑↓		↑		0	1/2		
↑↓		↓		0	−1/2		
↑↓			↑	−1	1/2		
↑↓			↓	−1	−1/2		

For a $2p^1$ occupancy:

p_1	p_0	p_{-1}	M_L	M_S		
↑			1	1/2	⇒	L = 1, S = 1/2
↓			1	−1/2		
	↑		0	1/2		
	↓		0	−1/2		
		↑	−1	1/2		
		↓	−1	−1/2		

Because the $1s^2$ occupancy represents a filled shell, it makes no net contribution to the angular momentum.

12. The s-orbitals are filled, and the two p-electrons are in different orbitals, and so are inequivalent. The angular momentum can be coupled by simple coupling of the individual angular momenta. Because both electrons are in p orbitals, L = 1 and S = 1/2. The possible values for L_{tot} are 2, 1, and 0; those for S_{tot} are 1 and 0. When coupling L_{tot} to $S_{tot} = 0$, the values of J are just equal to L_{tot}. This leads to term symbols of $^1D_2$, $^1P_1$, and $^1S_0$. When coupling the $S_{tot} = 1$ states to L_{tot}, there will be three values of $J = L_{tot} + 1$, L_{tot}, and $L_{tot} - 1$. These states lead to the following term symbols: $^3D_{3,2,1}$, $^3P_{2,1,0}$, and $^3S_{1,0}$.

13. For an oxygen atom with a $1s^2 2s^2 2p^4$ occupancy, the 1s and 2s electrons can be neglected, as filled shells have no net contribution to the total angular momentum. We need to construct a table of all possible arrangements of four electrons in a set of p-orbitals.

p_1	p_0	p_{-1}	M_L	M_S			
↑↓	↑↓		2	0 ⇒	L = 2, S = 0		
↑↓	↑	↑	1	1 ⇒		L = 1, S = 1	
↑↓	↑	↓	1	0	•		
↑↓	↓	↑	1	0		•	
↑↓	↓	↓	1	−1		•	
↑↓		↑↓	0	0	•		
↑	↑↓	↑	0	1		•	
↑	↑↓	↓	0	0		•	
↓	↑↓	↑	0	0 ⇒			L = 0, S = 0
↓	↑↓	↓	0	−1		•	
	↑↓	↑↓	−2	0	•		
↑	↑	↑↓	−1	1		•	
↑	↓	↑↓	−1	0	•		
↓	↑	↑↓	−1	0		•	
↓	↓	↑↓	−1	−1		•	

The individual states which contribute to the overall L = 2, S = 0 state are indicated by the first column of heavy dots. Likewise, those contributing to the L = 1 S = 1 are indicated by the second column of heavy dots. The term symbols from these three states are $^1D_2$, $^3P_{2,1,0}$, and $^1S_0$.

14. The ground state occupancy of the nitrogen atom is $1s^2 2s^2 2p^3$. As in the previous problem, we must make a table to find all the term symbols.

p_1	p_0	p_{-1}	M_L	M_S			
↑↓	↑		2	1/2	⇒ L = 2, S = 1/2		
↑↓	↓		2	−1/2	•		
↑↓		↑	1	1/2	•		
↑↓		↓	1	−1/2	•		
↑	↑	↑	0	3/2 ⇒		L = 0, S = 3/2	
↑	↑	↓	0	1/2	•		
↑	↓	↑	0	1/2		•	
↑	↓	↓	0	−1/2	•		
↓	↑	↑	0	1/2 ⇒			L = 0, S = 1/2
↓	↑	↓	0	−1/2		•	
↓	↓	↑	0	−1/2			•
↓	↓	↓	0	−3/2		•	
	↑↓	↑	−1	1/2	•		
	↑↓	↓	−1	−1/2	•		
	↑	↑↓	−2	1/2	•		
	↓	↑↓	−2	−1/2	•		

The term symbols for the nitrogen atom are $^2D_{5/2,3/2}$, $^4S_{3/2}$, and $^2S_{1/2}$. By the first of Hund's rules [p. 430], the ^4S state should be the ground state, as it has the highest spin multiplicity.

15. For a $3d^1$ occupancy, the table of spin states is

d_2	d_1	d_0	d_{-1}	d_{-2}	M_L	M_S	
↑					2	1/2	⇒ L = 2, S = 1/2
↓					2	−1/2	•
	↑				1	1/2	•
	↓				1	−1/2	•
		↑			0	1/2	•
		↓			0	−1/2	•
			↑		−1	1/2	•
			↓		−1	−1/2	•
				↑	−2	1/2	•
				↓	−2	−1/2	•

For a $3d^9$ occupancy, the table is

d_2	d_1	d_0	d_{-1}	d_{-2}	M_L	M_S	
↑	↑↓	↑↓	↑↓	↑↓	2	1/2	⇒ L = 2, S = 1/2
↓	↑↓	↑↓	↑↓	↑↓	2	−1/2	•
↑↓	↑	↑↓	↑↓	↑↓	1	1/2	•
↑↓	↓	↑↓	↑↓	↑↓	1	−1/2	•
↑↓	↑↓	↑	↑↓	↑↓	0	1/2	•
↑↓	↑↓	↓	↑↓	↑↓	0	−1/2	•
↑↓	↑↓	↑↓	↑	↑↓	−1	1/2	•
↑↓	↑↓	↑↓	↓	↑↓	−1	−1/2	•
↑↓	↑↓	↑↓	↑↓	↑	−2	1/2	•
↑↓	↑↓	↑↓	↑↓	↓	−2	−1/2	•

The resulting angular momentum state is the same for these two occupancies.

For the $3d^2$ occupancy:

d_2	d_1	d_0	d_{-1}	d_{-2}	M_L	M_S				
↑↓					4	0 ⇒	L = 4, S = 0			
↑	↑				3	1 ⇒		L = 3, S = 1		
↑	↓				3	0	•			
↓	↑				3	0		•		
↓	↓				3	−1		•		
↑		↑			2	1		•		
↑		↓			2	0	•			
↓		↑			2	0		•		
↓		↓			2	−1		•		
↑			↑		1	1		•		
↑			↓		1	0	•			
↓			↑		1	0		•		
↓			↓		1	−1		•		
↑				↑	0	1		•		
↑				↓	0	0	•			
↓				↑	0	0		•		
↓				↓	0	−1		•		
	↑↓				2	0 ⇒			L = 2, S = 0	
	↑	↑			1	1 ⇒				L = 1, S = 1
	↑	↓			1	0			•	

d_2	d_1	d_0	d_{-1}	d_{-2}	M_L	M_S					
	↓	↑			1	0					•
	↓	↓			1	–1					•
	↑		↑		0	1					•
	↑		↓		0	0				•	
	↓		↑		0	0					•
	↓		↓		0	–1					•
	↑			↑	–1	1			•		
	↑			↓	–1	0		•			
	↓			↑	–1	0			•		
	↓			↓	–1	–1			•		
		↑↓			0	0	⇒ L = 0, S = 0				
		↑	↑		–1	1					•
		↑	↓		–1	0				•	
		↓	↑		–1	0					•
		↓	↓		–1	–1					•
		↑		↑	–2	1			•		
		↑		↓	–2	0		•			
		↓		↑	–2	0			•		
		↓		↓	–2	–1			•		
			↑↓		–2	0				•	
			↑	↑	–3	1			•		
			↑	↓	–3	0		•			
			↓	↑	–3	0			•		
			↓	↓	–3	–1			•		
				↑↓	–4	0		•			

For the $3d^8$ occupancy:

d_2	d_1	d_0	d_{-1}	d_{-2}	M_L	M_S			
	↑↓	↑↓	↑↓	↑↓	–4	0	⇒ L = 4, S = 0		
↑	↑	↑↓	↑↓	↑↓	–3	1	⇒ L = 3, S = 1		
↑	↓	↑↓	↑↓	↑↓	–3	0		•	
↓	↑	↑↓	↑↓	↑↓	–3	0			•
↓	↓	↑↓	↑↓	↑↓	–3	–1			•
↑	↑↓	↑	↑↓	↑↓	–2	1			•
↑	↑↓	↓	↑↓	↑↓	–2	0		•	
↓	↑↓	↑	↑↓	↑↓	–2	0			•
↓	↑↓	↓	↑↓	↑↓	–2	–1			•

d_2	d_1	d_0	d_{-1}	d_{-2}	M_L	M_S				
↑	↑↓	↑↓	↑	↑↓	−1	1		•		
↑	↑↓	↑↓	↓	↑↓	−1	0	•			
↓	↑↓	↑↓	↑	↑↓	−1	0		•		
↓	↑↓	↑↓	↓	↑↓	−1	−1		•		
↑	↑↓	↑↓	↑↓	↑	−0	1		•		
↑	↑↓	↑↓	↑↓	↓	−0	0	•			
↓	↑↓	↑↓	↑↓	↑	−0	0		•		
↓	↑↓	↑↓	↑↓	↓	−0	−1		•		
↑↓		↑↓	↑↓	↑↓	−2	0 ⇒		L = 2, S = 0		
↑↓	↑	↑	↑↓	↑↓	−1	1 ⇒			L = 1, S = 1	
↑↓	↑	↓	↑↓	↑↓	−1	0			•	
↑↓	↓	↑	↑↓	↑↓	−1	0				•
↑↓	↓	↓	↑↓	↑↓	−1	−1				•
↑↓	↑	↑↓	↑	↑↓	0	1				•
↑↓	↑	↑↓	↓	↑↓	0	0			•	
↑↓	↓	↑↓	↑	↑↓	0	0				•
↑↓	↓	↑↓	↓	↑↓	0	−1				•
↑↓	↑	↑↓	↑↓	↑	1	1		•		
↑↓	↑	↑↓	↑↓	↓	1	0	•			
↑↓	↓	↑↓	↑↓	↑	1	0		•		
↑↓	↓	↑↓	↑↓	↓	1	−1		•		
↑↓	↑↓		↑↓	↑↓	0	0 ⇒	L = 0, S = 0			
↑↓	↑↓	↑	↑	↑↓	1	1				•
↑↓	↑↓	↑	↓	↑↓	1	0			•	
↑↓	↑↓	↓	↑	↑↓	1	0				•
↑↓	↑↓	↓	↓	↑↓	1	−1				•
↑↓	↑↓	↑	↑↓	↑	2	1		•		
↑↓	↑↓	↑	↑↓	↓	2	0	•			
↑↓	↑↓	↓	↑↓	↑	2	0		•		
↑↓	↑↓	↓	↑↓	↓	2	−1		•		
↑↓	↑↓	↑↓		↑↓	2	0			•	
↑↓	↑↓	↑↓	↑	↑	3	1		•		
↑↓	↑↓	↑↓	↑	↓	3	0	•			
↑↓	↑↓	↑↓	↓	↑	3	0		•		
↑↓	↑↓	↑↓	↓	↓	3	−1		•		
↑↓	↑↓	↑↓	↑↓		4	0	•			

This table is identical with that of the $3d^2$ occupancy, except the sign is changed for each value of M_L. However, any angular momentum state contains both $+M_L$ and $-M_L$ individual states; therefore, the term symbols of both tables are the same.

The pattern suggested by these two examples (which is true in general) is that the term symbols of a given electron occupancy are also those of an "electron vacancy." For a given number of electrons in a shell, the same term symbols will describe the state where the same numbers of electrons are missing from the shell.

16. The ground-state occupancy of Mn is $1s^2 2s^2 2p^6 3s^2 3p^6 4s^2 3d^5$.

For a table as lengthy as this one, a shorthand notation can greatly reduce the size of the table of possible occupations. Instead of indicating the spin state of each electron, we indicate their individual orbital occupancy with a vertical bar; then we write all the possible values of M_S for that entry.

There are only three different arrangements for five electrons in five equivalent orbitals: four electrons in two filled orbitals and a single unpaired electron, one pair of electrons and three unpaired, and all five unpaired.

<table>
<tr><th>d_2</th><th>d_1</th><th>d_0</th><th>d_{-1}</th><th>d_{-2}</th><th>M_L</th><th>M_S</th></tr>
<tr><td>||</td><td>||</td><td>|</td><td></td><td></td><td>6</td><td>±1/2</td></tr>
<tr><td>||</td><td>||</td><td></td><td>|</td><td></td><td>5</td><td>±1/2</td></tr>
<tr><td>||</td><td>||</td><td></td><td></td><td>|</td><td>4</td><td>±1/2</td></tr>
<tr><td>||</td><td>|</td><td>||</td><td></td><td></td><td>5</td><td>±1/2</td></tr>
<tr><td>||</td><td></td><td>||</td><td>|</td><td></td><td>3</td><td>±1/2</td></tr>
<tr><td>||</td><td></td><td>||</td><td></td><td>|</td><td>2</td><td>±1/2</td></tr>
<tr><td>||</td><td>|</td><td></td><td>||</td><td></td><td>3</td><td>±1/2</td></tr>
<tr><td>||</td><td></td><td>|</td><td>||</td><td></td><td>2</td><td>±1/2</td></tr>
<tr><td>||</td><td></td><td></td><td>||</td><td>|</td><td>0</td><td>±1/2</td></tr>
<tr><td>||</td><td>|</td><td></td><td></td><td>||</td><td>1</td><td>±1/2</td></tr>
<tr><td>||</td><td></td><td>|</td><td></td><td>||</td><td>0</td><td>±1/2</td></tr>
<tr><td>||</td><td></td><td></td><td>|</td><td>||</td><td>−1</td><td>±1/2</td></tr>
<tr><td>|</td><td>||</td><td>||</td><td></td><td></td><td>4</td><td>±1/2</td></tr>
<tr><td></td><td>||</td><td>||</td><td>|</td><td></td><td>1</td><td>±1/2</td></tr>
<tr><td></td><td>||</td><td>||</td><td></td><td>|</td><td>0</td><td>±1/2</td></tr>
<tr><td>|</td><td>||</td><td></td><td>||</td><td></td><td>2</td><td>±1/2</td></tr>
<tr><td></td><td>||</td><td>|</td><td>||</td><td></td><td>0</td><td>±1/2</td></tr>
</table>

<table>
<tr><th>d_2</th><th>d_1</th><th>d_0</th><th>d_{-1}</th><th>d_{-2}</th><th>M_L</th><th>M_S</th></tr>
<tr><td></td><td>||</td><td></td><td>||</td><td>|</td><td>−2</td><td>±1/2</td></tr>
<tr><td>|</td><td>||</td><td></td><td></td><td>||</td><td>0</td><td>±1/2</td></tr>
<tr><td></td><td>||</td><td>|</td><td></td><td>||</td><td>−2</td><td>±1/2</td></tr>
<tr><td></td><td>||</td><td></td><td>|</td><td>||</td><td>−3</td><td>±1/2</td></tr>
<tr><td>|</td><td></td><td>||</td><td>||</td><td></td><td>0</td><td>±1/2</td></tr>
<tr><td></td><td>|</td><td>||</td><td>||</td><td></td><td>−1</td><td>±1/2</td></tr>
<tr><td></td><td></td><td>||</td><td>||</td><td>|</td><td>−4</td><td>±1/2</td></tr>
<tr><td>|</td><td></td><td>||</td><td></td><td>||</td><td>−2</td><td>±1/2</td></tr>
<tr><td></td><td>|</td><td>||</td><td></td><td>||</td><td>−3</td><td>±1/2</td></tr>
<tr><td></td><td></td><td>||</td><td>|</td><td>||</td><td>−5</td><td>±1/2</td></tr>
<tr><td>|</td><td></td><td></td><td>||</td><td>||</td><td>−4</td><td>±1/2</td></tr>
<tr><td></td><td>|</td><td></td><td>||</td><td>||</td><td>−5</td><td>±1/2</td></tr>
<tr><td></td><td></td><td>|</td><td>||</td><td>||</td><td>−6</td><td>±1/2</td></tr>
<tr><td>||</td><td>|</td><td>|</td><td>|</td><td></td><td>4</td><td>±3/2, ±1/2, ±1/2, ±1/2</td></tr>
<tr><td>||</td><td>|</td><td>|</td><td></td><td>|</td><td>3</td><td>±3/2, ±1/2, ±1/2, ±1/2</td></tr>
<tr><td>||</td><td>|</td><td></td><td>|</td><td>|</td><td>2</td><td>±3/2, ±1/2, ±1/2, ±1/2</td></tr>
<tr><td>||</td><td></td><td>|</td><td>|</td><td>|</td><td>1</td><td>±3/2, ±1/2, ±1/2, ±1/2</td></tr>
<tr><td>|</td><td>||</td><td>|</td><td>|</td><td></td><td>3</td><td>±3/2, ±1/2, ±1/2, ±1/2</td></tr>
<tr><td>|</td><td>||</td><td>|</td><td></td><td>|</td><td>2</td><td>±3/2, ±1/2, ±1/2, ±1/2</td></tr>
<tr><td>|</td><td>||</td><td></td><td>|</td><td>|</td><td>1</td><td>±3/2, ±1/2, ±1/2, ±1/2</td></tr>
<tr><td></td><td>||</td><td>|</td><td>|</td><td>|</td><td>−1</td><td>±3/2, ±1/2, ±1/2, ±1/2</td></tr>
<tr><td>|</td><td>|</td><td>||</td><td>|</td><td></td><td>2</td><td>±3/2, ±1/2, ±1/2, ±1/2</td></tr>
<tr><td>|</td><td>|</td><td>||</td><td></td><td>|</td><td>1</td><td>±3/2, ±1/2, ±1/2, ±1/2</td></tr>
<tr><td>|</td><td></td><td>||</td><td>|</td><td>|</td><td>−1</td><td>±3/2, ±1/2, ±1/2, ±1/2</td></tr>
<tr><td></td><td>|</td><td>||</td><td>|</td><td>|</td><td>−2</td><td>±3/2, ±1/2, ±1/2, ±1/2</td></tr>
<tr><td>|</td><td>|</td><td>|</td><td>||</td><td></td><td>1</td><td>±3/2, ±1/2, ±1/2, ±1/2</td></tr>
<tr><td>|</td><td>|</td><td></td><td>||</td><td>|</td><td>−1</td><td>±3/2, ±1/2, ±1/2, ±1/2</td></tr>
<tr><td>|</td><td></td><td>|</td><td>||</td><td>|</td><td>−2</td><td>±3/2, ±1/2, ±1/2, ±1/2</td></tr>
<tr><td></td><td>|</td><td>|</td><td>||</td><td>|</td><td>−3</td><td>±3/2, ±1/2, ±1/2, ±1/2</td></tr>
<tr><td>|</td><td>|</td><td>|</td><td></td><td>||</td><td>−1</td><td>±3/2, ±1/2, ±1/2, ±1/2</td></tr>
<tr><td>|</td><td>|</td><td></td><td>|</td><td>||</td><td>−2</td><td>±3/2, ±1/2, ±1/2, ±1/2</td></tr>
<tr><td>|</td><td></td><td>|</td><td>|</td><td>||</td><td>−3</td><td>±3/2, ±1/2, ±1/2, ±1/2</td></tr>
<tr><td></td><td>|</td><td>|</td><td>|</td><td>||</td><td>−4</td><td>±3/2, ±1/2, ±1/2, ±1/2</td></tr>
<tr><td>|</td><td>|</td><td>|</td><td>|</td><td>|</td><td>0</td><td>±5/2, 5 × ±3/2, 10 × ±1/2</td></tr>
</table>

When all these terms are arranged to form angular momentum states, the term symbols for this occupation are found to be ^6S, ^4G, ^4F, ^4D, ^4P, ^2I, ^2H, ^2G, ^2G, ^2F, ^2F, ^2D, ^2D, ^2D, ^2P, ^2S.

Now, we repeat this for Mn^+, whose occupancy is $1s^22s^22p^63s^23p^64s^23d^4$. The three possible arrangements for four electrons in five d–orbitals are two sets of paired electrons, one set of paired electrons and two unpaired, or all four unpaired.

d_2	d_1	d_0	d_{-1}	d_{-2}	M_L	M_S
||	||				6	0
||		||			4	0
||			||		2	0
||				||	0	0
	||	||			2	0
	||		||		0	0
	||			||	–2	0
		||	||		–2	0
		||		||	–4	0
			||	||	–6	0
||	|	|			5	±1, 0, 0
||	|		|		4	±1, 0, 0
||	|			|	3	±1, 0, 0
||		|	|		3	±1, 0, 0
||		|		|	2	±1, 0, 0
||			|	|	1	±1, 0, 0
|	||	|			4	±1, 0, 0
|	||		|		3	±1, 0, 0
|	||			|	2	±1, 0, 0
	||	|	|		1	±1, 0, 0
	||	|		|	0	±1, 0, 0
	||		|	|	–1	±1, 0, 0
|	|	||			3	±1, 0, 0
|		||	|		1	±1, 0, 0
|		||		|	0	±1, 0, 0
	|	||	|		0	±1, 0, 0
	|	||		|	–1	±1, 0, 0
		||	|	|	–3	±1, 0, 0
|	|		||		1	±1, 0, 0
|		|	||		0	±1, 0, 0
|			||	|	–2	±1, 0, 0
	|	|	||		–1	±1, 0, 0

d_2	d_1	d_0	d_{-1}	d_{-2}	M_L	M_S
	$\vert$		$\Vert$	$\vert$	–3	±1, 0, 0
		$\vert$	$\Vert$	$\vert$	–4	±1, 0, 0
$\vert$	$\vert$			$\Vert$	–1	±1, 0, 0
$\vert$		$\vert$		$\Vert$	–2	±1, 0, 0
$\vert$			$\vert$	$\Vert$	–3	±1, 0, 0
	$\vert$	$\vert$		$\Vert$	–3	±1, 0, 0
	$\vert$		$\vert$	$\Vert$	–4	±1, 0, 0
		$\vert$	$\vert$	$\Vert$	–5	±1, 0, 0
$\vert$	$\vert$	$\vert$	$\vert$		2	±2, 4 × ±1, 6 × 0
$\vert$	$\vert$	$\vert$		$\vert$	1	±2, 4 × ±1, 6 × 0
$\vert$	$\vert$		$\vert$	$\vert$	0	±2, 4 × ±1, 6 × 0
$\vert$		$\vert$	$\vert$	$\vert$	–1	±2, 4 × ±1, 6 × 0
	$\vert$	$\vert$	$\vert$		–2	±2, 4 × ±1, 6 × 0

After collecting the individual states, we find the final set of term symbols to contain ^{5}D, ^{3}H, ^{3}G, ^{3}F, ^{3}F, ^{3}D, ^{3}P, ^{3}P, ^{1}I, ^{1}G, ^{1}G, ^{1}F, ^{1}D, ^{1}D, ^{1}S, and ^{1}S.

17. The occupancy of the valence shell of the halogens is p^5. As we saw in Problem 15, this occupancy has the same atomic states (term symbols) as p^1. The states for p^1 were worked out in Problem 11, and it was shown that $S = 1/2$, thus the ground state should be a doublet.

18. The angular momentum states for $3d^2$ were found in Problem 15. The terms symbols for these states are

$$L = 4, S = 0 \Rightarrow {}^1G_4.$$

$$L = 3, S = 1 \Rightarrow {}^3F_{4,3,2}.$$

$$L = 2, S = 0 \Rightarrow {}^1D_2.$$

$$L = 1, S = 1 \Rightarrow {}^3P_{2,1,0}.$$

$$E_{4,4,0} = (\gamma\hbar^2/2)[(4 \times 5) - (4 \times 5) - 0] = 0.$$

$$E_{4,3,1} = (\gamma\hbar^2/2)[(4 \times 5) - (3 \times 4) - (1 \times 2)] = 3\gamma\hbar^2.$$

$$E_{3,3,1} = (\gamma\hbar^2/2)[(3 \times 4) - (3 \times 4) - (1 \times 2)] = -\gamma\hbar^2.$$

$$E_{2,3,1} = (\gamma\hbar^2/2)[(2 \times 3) - (3 \times 4) - (1 \times 2)] = -4\gamma\hbar^2.$$

$E_{2,2,0} = (\gamma\hbar^2/2)[(2 \times 3) - (2 \times 3) - 0] = 0.$

$E_{2,1,1} = (\gamma\hbar^2/2)[(2 \times 3) - (1 \times 2) - (1 \times 2)] = \gamma\hbar^2.$

$E_{1,1,1} = (\gamma\hbar^2/2)[(1 \times 2) - (1 \times 2) - (1 \times 2)] = -\gamma\hbar^2.$

$E_{0,1,1} = (\gamma\hbar^2/2)[(0) - (1 \times 2) - (1 \times 2)] = -2\gamma\hbar^2.$

19. $E^{spin-orbit} = (\gamma\hbar^2/2)\,[\,J(J + 1) - L(L + 1) - S(S + 1)]$

For specific values of L and S, the maximum spin–orbit interaction energy will arise from the state of highest J value (since the J(J + 1) term enters the energy expression positively). The maximum value of J is L + S, so the maximum value of the spin–orbit interaction is

$$E^{spin-orbit}{}_{max} = (\gamma\hbar^2/2)\,[\,(L+S)(L + S + 1) - L(L + 1) - S(S + 1)]$$

$$= \gamma\hbar^2 LS.$$

20. The first step is to determine the ground state of the carbon atom. Its ground state occupancy is $1s^22s^22p^2$. We start by setting up a table of all possible states.

p_1	p_0	p_{-1}	M_L	M_S		
↑↓			2	0	⇒ L = 2, S = 0	
	↑↓		0	0	•	
		↑↓	–2	0	•	
↑↓	↑	↑	1	1	⇒	L = 1, S = 1
↑↓	↑	↓	1	0	•	
↑↓	↓	↑	1	0		•
↑↓	↓	↓	1	–1		•
↑	↑↓	↑	0	1		•
↑	↑↓	↓	0	0	•	
↓	↑↓	↑	0	0		•
↓	↑↓	↓	0	–1		•
↑	↑	↑↓	–1	1		•
↑	↓	↑↓	–1	0	•	
↓	↑	↑↓	–1	0		•
↓	↓	↑↓	–1	–1		•

The term symbols for the ground state of carbon are ^3P and ^1D. ^3P will be the ground state by Hund's rule (it has the greater spin multiplicity).

For the excited-state occupancy, $1s^2 2s^1 2p^3$, the single s electron in the 2s orbital will have L = 0 and S = 1/2. The states of the three p electrons will have to be determined.

p_1	p_0	p_{-1}	M_L	M_S				
↑↓	↑		2	1/2 ⇒	L = 2, S = 1/2			
↑↓	↓		2	−1/2	•			
↑↓		↑	1	1/2	•			
↑↓		↓	1	−1/2	•			
↑	↑↓		1	1/2 ⇒		L = 1, S = 1/2		
↓	↑↓		1	−1/2		•		
	↑↓	↑	−1	1/2	•			
	↑↓	↓	−1	−1/2	•			
↑		↑↓	−1	1/2		•		
↓		↑↓	−1	−1/2		•		
	↑	↑↓	−2	1/2	•			
	↓	↑↓	−2	−1/2	•			
↑	↑	↑	0	3/2 ⇒			L = 0, S = 3/2	
↓	↑	↑	0	1/2		•		
↑	↓	↑	0	1/2			•	
↑	↑	↓	0	1/2 ⇒				L = 0, S = 1/2
↑	↓	↓	0	−1/2		•		
↓	↑	↓	0	−1/2			•	
↓	↓	↑	0	−1/2				•
↓	↓	↓	0	−3/2			•	

The states of the p^3 occupancy now must be coupled to the s^1 electron. This leads to states of L = 2, S = 1,0 (^3D, ^1D), L = 1, S = 1,0 (^3P, ^1P), L = 0, S = 2,1 (^5S, ^3S) and L = 0, S = 1,0 (^3S, ^1S).

The selection rules for many-electron atoms are that $|\Delta L| = 1$ and $\Delta S = 0$ [10-26]. The ground state is ^3P (L = 1, S = 1), so transitions are allowed to the ^3S ($\Delta L = -1$, $\Delta S = 0$) and ^3D ($\Delta L = +1$, $\Delta S = 0$) states.

21. For neon, whose ground state occupancy is $1s^22s^22p^6$, the term symbol is $^1S_0$, since the all the orbital shells are filled. The selection rules for allowed transitions require that one and only one electron change its l quantum number at a time, and only by ±1. So, the first excited occupancy must be $1s^22s^22p^53s^1$.

    This will not require a table procedure, because as we've seen, the angular momentum of a p^5 occupancy is the same as that of a p^1 occupancy, which is just L = 1, S = 1/2. The s electron has L = 0, S = 1/2. Now, these inequivalent electrons can be simply coupled together, with the final angular momentum states being L = 1, S = 1,0; the associated term symbols are ^3P and ^1P.

    The selection rules are $\Delta S = 0$ and $\Delta L = \pm 1$ [10-26]; thus the allowed transition is from the ground-state ^1S to the ^1P excited state.

22. The angular momentum states of a $1s^22s^22p^2$ occupancy were worked out in Problem 20, as this is the ground state of the carbon atom. The L,S states were found to be L = 2, S = 0 and L = 1, S = 1. These states can be coupled with the L = 2, S = 1/2 state of the d^1 electron. The final states are L = 4,3,2,1,0, S = 1/2; L = 3, S = 3/2,1/2; L = 2, S = 3/2,1/2; and L = 1, S = 3/2,1/2. The term symbols associated with these states are ^2G, ^2F, ^2D, ^2P, ^2S, ^4F, ^2F, ^4D, ^2D, ^4P, and ^2P.

23. A spatial orbital could contain four spin–3/2 particles and still not have any particles with identical quantum numbers, since the z–component of the spin could be 3/2, 1/2, –1/2, or –3/2. Thus, there are four different spin–orbitals to combine with a given spatial orbital.

24. $\Phi = \phi_1(1)\,\phi_2(2)\,\phi_3(3)\,\phi_4(4)$.

    $\Phi_{12} = (1 - P_{12})\,\Phi = (1/2^{1/2})[\phi_1(1)\,\phi_2(2)\,\phi_3(3)\,\phi_4(4) - \phi_1(2)\,\phi_2(1)\,\phi_3(3)\,\phi_4(4)]$.

    $\Phi_{123} = (1 - P_{13} - P_{23})\Phi_{12} = (1/6^{1/2})[\phi_1(1)\,\phi_2(2)\,\phi_3(3)\,\phi_4(4)$
    $- \phi_1(2)\,\phi_2(1)\,\phi_3(3)\,\phi_4(4) - \phi_1(3)\,\phi_2(2)\,\phi_3(1)\,\phi_4(4) + \phi_1(2)\,\phi_2(3)\,\phi_3(1)\,\phi_4(4)$
    $- \phi_1(1)\,\phi_2(3)\,\phi_3(2)\,\phi_4(4) + \phi_1(3)\,\phi_2(1)\,\phi_3(2)\,\phi_4(4)]$.

$$\Phi_{1234} = (1 - P_{14} - P_{24} - P_{34})\Phi_{123} = (1/24^{1/2})[\phi_1(1)\,\phi_2(2)\,\phi_3(3)\,\phi_4(4)$$
$$- \phi_1(2)\,\phi_2(1)\,\phi_3(3)\,\phi_4(4) - \phi_1(3)\,\phi_2(2)\,\phi_3(1)\,\phi_4(4) + \phi_1(2)\,\phi_2(3)\,\phi_3(1)\,\phi_4(4)$$
$$- \phi_1(1)\,\phi_2(3)\,\phi_3(2)\,\phi_4(4) + \phi_1(3)\,\phi_2(1)\,\phi_3(2)\,\phi_4(4) - \phi_1(4)\,\phi_2(2)\,\phi_3(3)\,\phi_4(1)$$
$$+ \phi_1(2)\,\phi_2(4)\,\phi_3(3)\,\phi_4(1) + \phi_1(3)\,\phi_2(2)\,\phi_3(4)\,\phi_4(1) - \phi_1(2)\,\phi_2(3)\,\phi_3(4)\,\phi_4(1)$$
$$+ \phi_1(4)\,\phi_2(3)\,\phi_3(2)\,\phi_4(1) - \phi_1(3)\,\phi_2(4)\,\phi_3(2)\,\phi_4(1) - \phi_1(1)\,\phi_2(4)\,\phi_3(3)\,\phi_4(2)$$
$$+ \phi_1(4)\,\phi_2(1)\,\phi_3(3)\,\phi_4(2) + \phi_1(3)\,\phi_2(4)\,\phi_3(1)\,\phi_4(2) - \phi_1(4)\,\phi_2(3)\,\phi_3(1)\,\phi_4(2)$$
$$+ \phi_1(1)\,\phi_2(3)\,\phi_3(4)\,\phi_4(2) - \phi_1(3)\,\phi_2(1)\,\phi_3(4)\,\phi_4(2) - \phi_1(1)\,\phi_2(2)\,\phi_3(4)\,\phi_4(3)$$
$$+ \phi_1(2)\,\phi_2(1)\,\phi_3(4)\,\phi_4(3) + \phi_1(4)\,\phi_2(2)\,\phi_3(1)\,\phi_4(3) - \phi_1(2)\,\phi_2(4)\,\phi_3(1)\,\phi_4(3)$$
$$+ \phi_1(1)\,\phi_2(4)\,\phi_3(2)\,\phi_4(3) - \phi_1(4)\,\phi_2(1)\,\phi_3(2)\,\phi_4(3)].$$

Now, we can compare this with the function produced by the Slater determinant.

$$\Phi_{1234} = \frac{1}{\sqrt{24}}\begin{vmatrix} \phi_1(1) & \phi_2(1) & \phi_3(1) & \phi_4(1) \\ \phi_1(2) & \phi_2(2) & \phi_3(2) & \phi_4(2) \\ \phi_1(3) & \phi_2(3) & \phi_3(3) & \phi_4(3) \\ \phi_1(4) & \phi_2(4) & \phi_3(4) & \phi_4(4) \end{vmatrix}$$

$$= (24^{-1/2})[\phi_1(1)\phi_2(2)\phi_3(3)\phi_4(4) - \phi_1(1)\phi_2(2)\phi_3(4)\phi_4(3) - \phi_1(1)\phi_2(3)\phi_3(2)\phi_4(4)$$
$$+ \phi_1(1)\phi_2(4)\phi_3(2)\phi_4(3) + \phi_1(1)\phi_2(3)\phi_3(4)\phi_4(2) - \phi_1(1)\phi_2(4)\phi_3(3)\phi_4(2)$$
$$- \phi_1(2)\phi_2(1)\phi_3(3)\phi_4(4) + \phi_1(2)\phi_2(1)\phi_3(4)\phi_4(3) + \phi_1(3)\phi_2(1)\phi_3(2)\phi_4(4)$$
$$- \phi_1(4)\phi_2(1)\phi_3(2)\phi_4(3) - \phi_1(3)\phi_2(1)\phi_3(4)\phi_4(2) + \phi_1(4)\phi_2(1)\phi_3(3)\phi_4(2)$$
$$+ \phi_1(2)\phi_2(3)\phi_3(1)\phi_4(4) - \phi_1(2)\phi_2(4)\phi_3(1)\phi_4(3) - \phi_1(3)\phi_2(2)\phi_3(1)\phi_4(4)$$
$$+ \phi_1(4)\phi_2(2)\phi_3(1)\phi_4(3) + \phi_1(3)\phi_2(4)\phi_3(1)\phi_4(2) - \phi_1(4)\phi_2(3)\phi_3(1)\phi_4(2)$$
$$- \phi_1(2)\phi_2(3)\phi_3(4)\phi_4(1) + \phi_1(2)\phi_2(4)\phi_3(3)\phi_4(1) + \phi_1(3)\phi_2(2)\phi_3(4)\phi_4(1)$$
$$- \phi_1(4)\phi_2(2)\phi_3(3)\phi_4(1) - \phi_1(3)\phi_2(4)\phi_3(2)\phi_4(1) + \phi_1(4)\phi_2(3)\phi_3(2)\phi_4(1)].$$

We can verify term by term that these two functions are identical.

25. $$\Phi = \frac{1}{\sqrt{6}}\begin{vmatrix} 1s\alpha(1) & 1s\alpha(1) & 1s\beta(1) \\ 1s\alpha(2) & 1s\alpha(2) & 1s\beta(2) \\ 1s\alpha(3) & 1s\alpha(3) & 1s\beta(3) \end{vmatrix}$$

$$= (6^{-1/2})\,[1s\alpha(1)1s\alpha(2)1s\beta(3) - 1s\alpha(1)1s\alpha(3)1s\beta(2)$$

$$- 1s\alpha(2)1s\alpha(1)1s\beta(3) + 1s\alpha(3)1s\alpha(1)1s\beta(2)$$

$$+ 1s\alpha(2)1s\alpha(3)1s\beta(1) - 1s\alpha(3)1s\alpha(2)1s\beta(1)\,]\,.$$

By inspecting the various terms, we see that terms 1 and 3 cancel each other, as do terms 2 and 4 and 5 and 6. The net result is that $\Phi = 0$.

26. $$\Phi = \frac{1}{\sqrt{6}}\begin{vmatrix} 1s\alpha(1) & 2s\alpha(1) & 3s\alpha(1) \\ 1s\alpha(2) & 2s\alpha(2) & 3s\alpha(2) \\ 1s\alpha(3) & 2s\alpha(3) & 3s\alpha(3) \end{vmatrix}$$

$$= (6^{-1/2})\,[1s\alpha(1)2s\alpha(2)3s\alpha(3) - 1s\alpha(1)2s\alpha(3)3s\alpha(2)$$

$$- 1s\alpha(2)2s\alpha(1)3s\alpha(3) + 1s\alpha(3)2s\alpha(1)3s\alpha(2)$$

$$+ 1s\alpha(2)2s\alpha(3)3s\alpha(1) - 1s\alpha(3)2s\alpha(2)3s\alpha(1)\,]\,.$$

$$(6^{-1/2})(1 - P_{13} - P_{23})(1 - P_{12})[1s\alpha(1)2s\alpha(2)3s\alpha(3)]$$

$$= (6^{-1/2})\,(1 - P_{13} - P_{23})\,[1s\alpha(1)2s\alpha(2)3s\alpha(3) - 1s\alpha(2)2s\alpha(1)3s\alpha(3)]$$

$$= (6^{-1/2})\,[1s\alpha(1)2s\alpha(2)3s\alpha(3) - 1s\alpha(2)2s\alpha(1)3s\alpha(3) - 1s\alpha(3)2s\alpha(2)3s\alpha(1)$$
$$+1s\alpha(2)2s\alpha(3)3s\alpha(1) - 1s\alpha(1)2s\alpha(3)3s\alpha(2) + 1s\alpha(3)2s\alpha(1)3s\alpha(2)].$$

$$(6^{-1/2})(1 - P_{13} - P_{23})(1 - P_{12})[-1s\alpha(1)2s\alpha(3)3s\alpha(2)]$$

$$= (6^{-1/2})\,(1 - P_{13} - P_{23})\,[-1s\alpha(1)2s\alpha(3)3s\alpha(2) + 1s\alpha(2)2s\alpha(3)3s\alpha(1)]$$

$$= (6^{-1/2})\,[-1s\alpha(1)2s\alpha(3)3s\alpha(2) + 1s\alpha(2)2s\alpha(3)3s\alpha(1) + 1s\alpha(3)2s\alpha(1)3s\alpha(2)$$
$$-1s\alpha(2)2s\alpha(1)3s\alpha(3) + 1s\alpha(1)2s\alpha(2)3s\alpha(3) - 1s\alpha(3)2s\alpha(2)3s\alpha(1)].$$

$$(6^{-1/2})(1 - P_{13} - P_{23})(1 - P_{12})[-1s\alpha(2)2s\alpha(1)3s\alpha(3)]$$

$= (6^{-1/2})\ (1 - P_{13} - P_{23})\ [-1s\alpha(2)2s\alpha(1)3s\alpha(3) + 1s\alpha(1)2s\alpha(2)3s\alpha(3)]$

$= (6^{-1/2})\ [-1s\alpha(2)2s\alpha(1)3s\alpha(3) + 1s\alpha(1)2s\alpha(2)3s\alpha(3) + 1s\alpha(2)2s\alpha(3)3s\alpha(1)$
$-1s\alpha(3)2s\alpha(2)3s\alpha(1) + 1s\alpha(3)2s\alpha(1)3s\alpha(2) - 1s\alpha(1)2s\alpha(3)3s\alpha(2)].$

$(6^{-1/2})(1 - P_{13} - P_{23})(1 - P_{12})[\ 1s\alpha(3)2s\alpha(1)3s\alpha(2)]$

$= (6^{-1/2})\ (1 - P_{13} - P_{23})\ [1s\alpha(3)2s\alpha(1)3s\alpha(2) - 1s\alpha(3)2s\alpha(2)3s\alpha(1)]$

$= (6^{-1/2})\ [1s\alpha(3)2s\alpha(1)3s\alpha(2) - 1s\alpha(3)2s\alpha(2)3s\alpha(1) - 1s\alpha(1)2s\alpha(3)3s\alpha(2)$
$+1s\alpha(1)2s\alpha(2)3s\alpha(3) - 1s\alpha(2)2s\alpha(1)3s\alpha(3) + 1s\alpha(2)2s\alpha(3)3s\alpha(1)].$

$(6^{-1/2})(1 - P_{13} - P_{23})(1 - P_{12})[\ 1s\alpha(2)2s\alpha(3)3s\alpha(1)]$

$= (6^{-1/2})\ (1 - P_{13} - P_{23})\ [1s\alpha(2)2s\alpha(3)3s\alpha(1) - 1s\alpha(1)2s\alpha(3)3s\alpha(2)]$

$= (6^{-1/2})\ [1s\alpha(2)2s\alpha(3)3s\alpha(1) - 1s\alpha(1)2s\alpha(3)3s\alpha(2) - 1s\alpha(2)2s\alpha(1)3s\alpha(3)$
$+1s\alpha(3)2s\alpha(1)3s\alpha(2) - 1s\alpha(3)2s\alpha(2)3s\alpha(1) + 1s\alpha(1)2s\alpha(2)3s\alpha(3)].$

$(6^{-1/2})(1 - P_{13} - P_{23})(1 - P_{12})[-\ 1s\alpha(3)2s\alpha(2)3s\alpha(1)]$

$= (6^{-1/2})\ (1 - P_{13} - P_{23})\ [-1s\alpha(3)2s\alpha(2)3s\alpha(1) + 1s\alpha(3)2s\alpha(1)3s\alpha(2)]$

$= (6^{-1/2})\ [-1s\alpha(3)2s\alpha(2)3s\alpha(1) + 1s\alpha(3)2s\alpha(1)3s\alpha(2) + 1s\alpha(1)2s\alpha(2)3s\alpha(3)$
$-1s\alpha(1)2s\alpha(3)3s\alpha(2) + 1s\alpha(2)2s\alpha(3)3s\alpha(1) - 1s\alpha(2)2s\alpha(1)3s\alpha(3)].$

In each case, applying A_N to each term generates all the other terms. If each of these resulting functions is added together, this represents applying A_N to Φ; Φ was formed by applying A_N to $1s\alpha 2s\alpha 3s\alpha$. This illustrates the idempotency of A_N: when applied to an antisymmetrized wavefunction, A_N has no further affect.

27. The following orbital correlation diagrams give a qualitative picture of the bonding for each of these molecules. The bond order for each is given under each diagram.

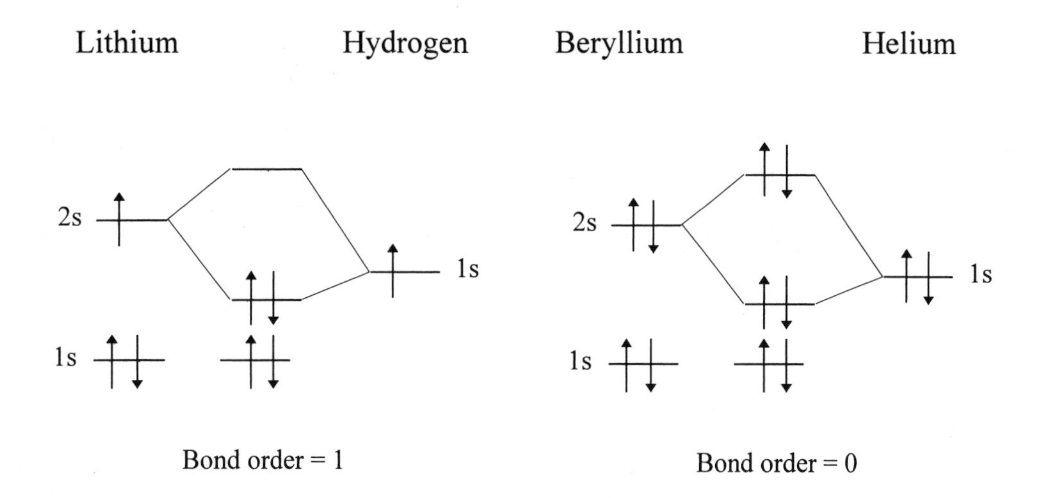

For LiH, the bond order is one, and the LCAO approximation predicts it will be stable. For BeHe, whose bond order is zero, LCAO predicts it to be unstable.

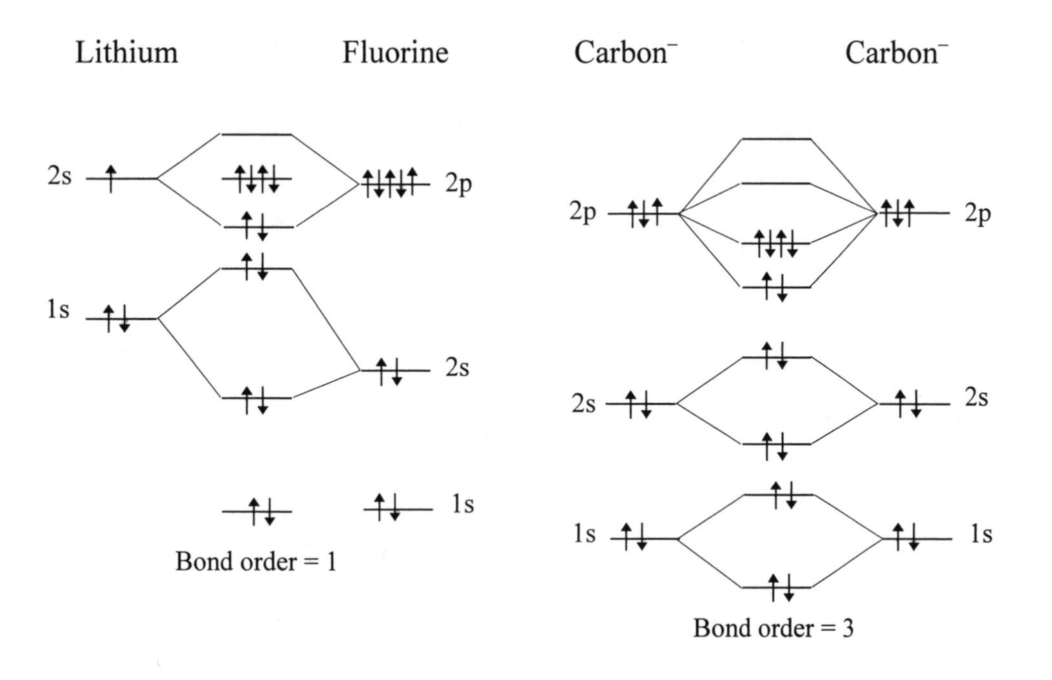

The bond orders of LiF and C_2^{2-} are one and 3, respectively. Both are predicted to be stable.

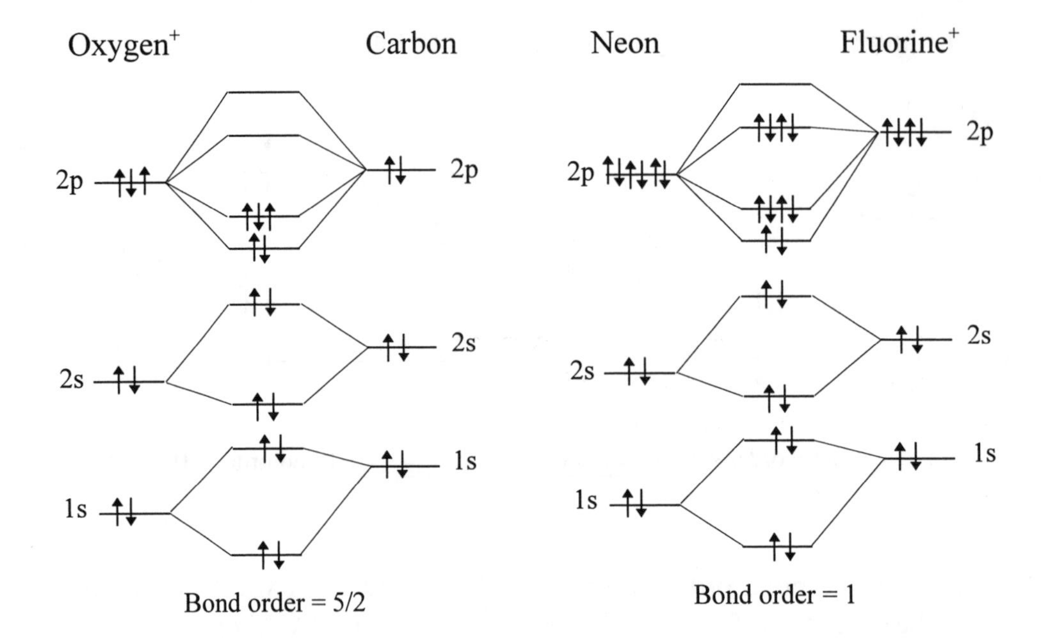

The bond orders of CO^+ and NeF^+ are 5/2 and 1; both are predicted to be stable.

28. Returning to [10-61], we have

$$\Delta E = T_e + (n'' + 1/2)\hbar\omega'' + B''J''(J'' + 1) - D''[J''(J'' + 1)]^2$$
$$- (n' + 1/2)\hbar\omega' + B'J'(J' + 1) + D'[J'(J' + 1)]^2.$$

From this general energy expression, we can now find the rotational contribution to ΔE.

$\Delta J = J'' - J' = 1$ (P–branch):

$$\Delta E^{rot} = B''(J' + 1)(J' + 2) - D''[(J' + 1)(J' + 2)]^2 - (B'J'(J' + 1) - D''[J'(J' + 1)]^2)$$

$$= -(D'' - D')\, J'^4 - (6D'' - 2D')\, J'^3 + (B'' - 13D'' - B' + D')J'^2$$
$$+ (3B'' - 12\, D'' - B')J' + 2B'' - 4D''.$$

$\Delta J = J'' - J' = 0$ (Q–branch):

$$\Delta E^{rot} = B''J'\,(J' + 1) + D''[J'(J' + 1)]^2 - (B'J'(J' + 1) - D''[J'(J' + 1)]^2)$$

$$= (B'' - B')\, J'\, (J' + 1) \; - (D'' - D')\, [J'(J' + 1)]^2.$$

$\Delta J = J'' - J' = -1$ (R–branch):

$$\Delta E^{rot} = B''(J' - 1)(J') - D''[(J' - 1)J']^2 - (B'J'(J' + 1) - D''[J'(J' + 1)]^2)$$

$$= -(D'' - D')J'^4 + 2(D'' + D')J'^3 + (B'' - D'' - B' + D')J'^2 - 2(B'' - B')J'.$$

29. For N_2, by symmetry it can never have a dipole moment, therefore vibrational transitions are forbidden, and IR spectroscopy cannot extract information about the molecule. For electronic transitions, the selection rules allow transitions between vibrational–rotational states of the different electronic states, and so information about N_2 can be extracted.

30. We have already dealt with the calculation of relative populations, in Problem 17 from Chapter 1. We can follow that example, only substituting in the rotational constant of LiH, 7.51 cm^{-1}.

$$E_J = 7.51\ cm^{-1} \times J \times (J+1).$$
$$g_J = (2 \times J) + 1.$$
$$\frac{P_J}{P_0} = \frac{(2J + 1) \times e^{-7.51 \times J(J+1)/208.3}}{1 \times e^{-0}} = (2J + 1)\ e^{-7.51 \times J(J+1)/208.3}.$$

The populations for J = 1 to 8 are

J:	1	2	3	4	5	6	7	8
P_J:	2.791	4.026	4.539	4.372	3.724	2.853	1.986	1.263

Now, we can sketch the transitions with these relative intensities for the two cases worked in Example 10.1

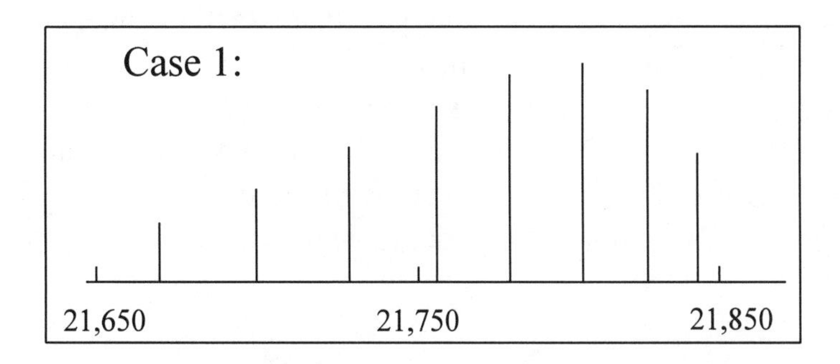

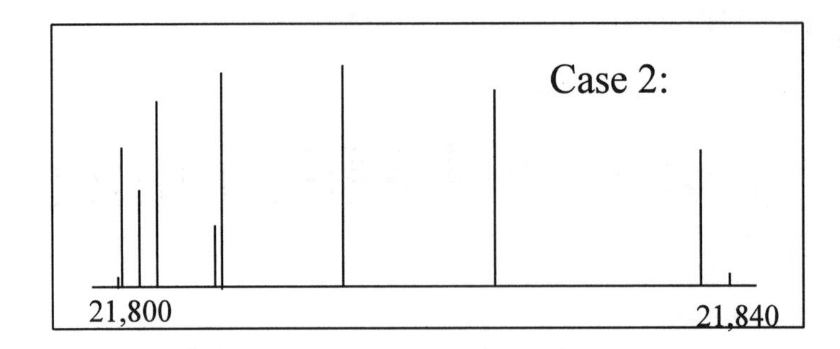

The qualitative difference in the appearance of the two spectra at low resolution would be the long tail in Case 1 compared with the abrupt band head in Case 2.

31. Note the example charge distribution on p. 463, where a –2 charge lies at the origin and two +1 charges lie equidistant from the negative charge. This is an example of a charge distribution that possesses no net charge or dipole moment but does have a quadrupole moment. An arrangement that likewise possesses no net charge, dipole, or quadrupole moments is

where the charges on each particle are listed, and the large particles lie at a distance d from the origin, and the small particles lie at a distance of 3d from the origin.

This arrangement can be thought of as two dipoles, one from the smaller charges and one from the larger charges, that exactly cancel each other out, leaving no net dipole moment. In the case of CO, there are likewise two distinct dipoles, one from the positive charges of the nuclei, and one from the negative charges of the electron. They nearly cancel one another out, leaving a very small net dipole. Yet, as in the distribution depicted above, both dipoles contribute to a large octupole moment.

Chapter 11

Statistical Mechanics

Exercises

1. For a $2s^1 2p^2$ occupancy of three electrons, there are two energy levels, 2s and 2p. We designate the 2s orbital as state 1 and the 2p orbital as state 2. Then, the degeneracies are $s_1 = 2$ and $s_2 = 6$, and the populations are $n_1 = 1$ and $n_2 = 2$.

   The number of ways the electrons can be distributed can be found using the table method introduced in the last chapter. The one electron in the s orbital has only two possibilities, $M_S = \pm 1/2$. The two p electrons can be arranged as a pair in one p orbital, or as two unpaired electrons.

p_1	p_0	p_{-1}	M_L	M_S
||			2	0
	||		0	0
		||	–2	0
|	|		1	±1, 0, 0
|		|	0	±1, 0, 0
	|	|	–1	±1, 0, 0

From this table we see that there are 15 ways to distribute the two electrons in the p orbital shell. Therefore, the total number of arrangements of the three electrons is $2 \times 15 = 30$.

Now, we apply [11-6]:

$$A_{\text{indistinguishable}} = \left(\frac{(2)}{1!}\right)\left(\frac{(6)(5)}{2!}\right) = 2 \times 15 = 30.$$

If the electrons were distinguishable, there would be three times as many ways for there to be one electron in the s orbital, for a total of 6 possibilities, and there would be two times as many arrangements of the two electrons in the p orbitals (e.g., if electron 1 was in the s orbital, the two electrons in the p orbital could be arranged as 2-3 or 3-2). Thus, the total number of arrangements would be $(3 \times 2) \times (2 \times 15) = 180$. The number of arrangements is increased by a factor of N! ($3! = 6$).

2. Nuclei with an odd number of protons + neutrons are fermions: ^{7}Li, ^{13}C, and ^{17}O. Nuclei with an even number of protons + neutrons are bosons: ^{6}Li, ^{12}C, ^{16}O, ^{18}O.

3. It is convenient to use the value 0.695 cm^{-1} K^{-1} for Boltzmann's constant.

$$q_v(1\ K) = \frac{e^{-1405\ cm^{-1}/(2\ \times\ 0.695\ cm^{-1}\ K^{-1}\ \times\ 1\ K)}}{1 - e^{-1405\ cm^{-1}/(0.695\ cm^{-1}\ K^{-1}\ \times\ 1\ K)}} = 1.04 \times 10^{-439}.$$

$$q_v(10\ K) = \frac{e^{-1405\ cm^{-1}/(2\ \times\ 0.695\ cm^{-1}\ K^{-1}\ \times\ 10\ K)}}{1 - e^{-1405\ cm^{-1}/(0.695\ cm^{-1}\ K^{-1}\ \times\ 10\ K)}} = 1.26 \times 10^{-44}.$$

$$q_v(100\ K) = \frac{e^{-1405\ cm^{-1}/(2\ \times\ 0.695\ cm^{-1}\ K^{-1}\ \times\ 100\ K)}}{1 - e^{-1405\ cm^{-1}/(0.695\ cm^{-1}\ K^{-1}\ \times\ 100\ K)}} = 4.08 \times 10^{-5}.$$

$$q_v(300\ K) = \frac{e^{-1405\ cm^{-1}/(2\ \times\ 0.695\ cm^{-1}\ K^{-1}\ \times\ 300\ K)}}{1 - e^{-1405\ cm^{-1}/(0.695\ cm^{-1}\ K^{-1}\ \times\ 300\ K)}} = 0.0345.$$

$$q_v(1000\text{ K}) = \frac{e^{-1405\text{ cm}^{-1}/(2\times 0.695\text{ cm}^{-1}\text{ K}^{-1}\times 1000\text{ K})}}{1-e^{-1405\text{ cm}^{-1}/(0.695\text{ cm}^{-1}\text{ K}^{-1}\times 1000\text{ K})}} = 0.419.$$

4. The probability that a given state is populated is given by [11-36]

$$P_i = e^{-E_i/kT}/q,$$

where q is the vibrational partition function, which is given by [11-46].

Since the vibrational energy is given in kJ mol^{-1}, it will be convenient to realize that Boltzmann's constant in J mol^{-1} is just the gas constant in joules, R = 8.314 J K^{-1} mol^{-1}, and so k = 8.314 × 10^{-3} kJ K^{-1} mol^{-1}:

$$P_2 = \frac{e^{-E_2/kT}}{\left(\dfrac{e^{-E_0/kT}}{1-e^{-2E_0/kT}}\right)} = \frac{e^{-E_2/kT}(1-e^{-2E_0/kT})}{e^{-E_0/kT}} = e^{-(E_2-E_0)/kT} - e^{-(E_2+E_0)/kT}.$$

$$P_2 = e^{-46\text{ kJ mol}^{-1}/(8.314\times 10^{-3}\text{ kJ mol}^{-1}\times T\text{ K})} - e^{-69\text{ kJ mol}^{-1}/(8.314\times 10^{-3}\text{ kJ mol}^{-1}\times T\text{ K})}.$$

$$P_2(100) = e^{-46/(8.314\times 10^{-3}\times 100)} - e^{-69\text{ kJ mol}^{-1}/(8.314\times 10^{-3}\times 100)} = 9.36\times 10^{-25}.$$

$$P_2(300) = e^{-46/(8.314\times 10^{-3}\times 300)} - e^{-69\text{ kJ mol}^{-1}/(8.314\times 10^{-3}\times 300)} = 9.78\times 10^{-9}.$$

$$P_2(1000) = e^{-46/(8.314\times 10^{-3}\times 1000)} - e^{-69\text{ kJ mol}^{-1}/(8.314\times 10^{-3}\times 1000)} = 0.00371.$$

$$P_2(10000) = e^{-46/(8.314\times 10^{-3}\times 10000)} - e^{-69\text{ kJ mol}^{-1}/(8.314\times 10^{-3}\times 1000)} = 0.139.$$

5. $e^{-E/kT} = 0.001.$

$-E/kT = \ln(0.001) = -6.907.$

T = E kJ mol^{-1} / (8.314 × 10^{-3} kJ mol^{-1} × 6.907).

T = 1.0 kJ mol^{-1} / (8.314 × 10^{-3} kJ mol^{-1} × 6.907) = 17.41 K.

T = 300.0 kJ mol^{-1} / (8.314 × 10^{-3} kJ mol^{-1} × 6.907) = 5223 K.

$$T = 10^5 \text{ kJ mol}^{-1} / (8.314 \times 10^{-3} \text{ kJ mol}^{-1} \times 6.907) = 1.741 \times 10^6 \text{ K}.$$

Because the temperatures at which the typical first excited state's population is only 0.001 (0.1%) are much greater than room temperature, it is appropriate to neglect any population of the excited electronic or nuclear states.

Additional Exercises

6. For a classical one-dimensional harmonic oscillator, the total energy is given by the sum of the kinetic energy and the potential energy.

$$E = p^2/2m + 1/2\ kx^2$$

The phase space variables are x and p, so the trajectory consists of all x,p values that give the total energy; this trajectory is simply an ellipse. It intersects the x-axis at $x = \pm\ (2E/k\ )^{1/2}$ [when p = 0, the kinetic energy = 0, so all the energy has to be potential energy] and the p-axis at $p = \pm(2mE)^{1/2}$.

If the energy is doubled, the shape of the ellipse remains the same but is expanded along both axes. If the mass is doubled while the energy is held constant, the trajectory will be extended by a factor of $2^{1/2}$ along the p-axis.

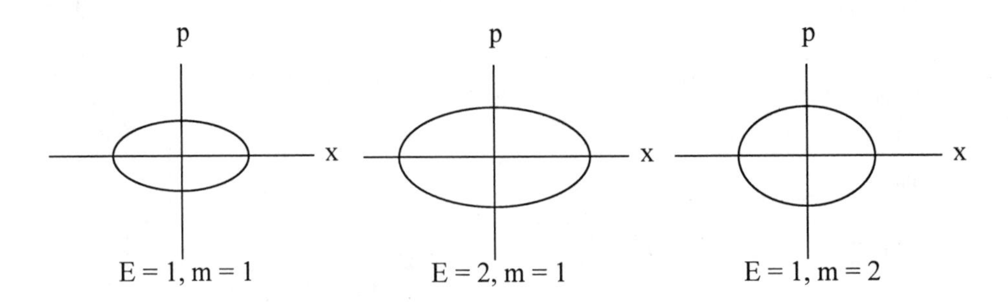

7. Equation [11-10] provides a general formula for the number of arrangements of N particles where n_i of the particles are in state i with a degeneracy of s_i:

$$A = \prod_i \frac{s_i(s_i - \delta)(s_i - 2\delta)(s_i - 3\delta) \cdots (s_i - (n-1)\delta)}{n_i!}.$$

If we now substitute in $\delta = 1$ for fermions, the equation becomes

$$A = \prod_i \frac{s_i(s_i - 1)(s_i - 2)(s_i - 3) \cdots (s_i - (n-1))}{n_i!}.$$

We recall the definition of s!:

$$s! = s\,(s-1)\,(s-2) \cdots (s-(n-1))\,(s-n)\,(s-(n+1)) \cdots (2)\,(1).$$

It should be apparent that the terms beginning with (s – n) and ending with (1) are simply (s – n)!, which leads to

$$s!\,/\,(s-n)! = s\,(s-1)\,(s-2) \cdots (s-(n-1)).$$

We can now substitute the two factorial terms back into the expression above to arrive at [11-11]:

$$A = \prod_i \frac{(s_i!\ /\ (s-n)!)}{n_i!} = \prod_i \frac{s_i!}{n_i!\,(s-n)!}.$$

8.

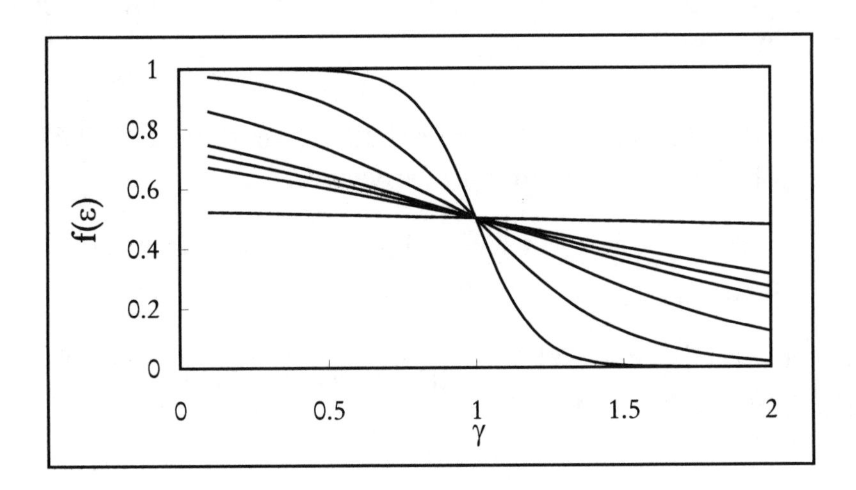

In the figure above, each curve represents a different temperature; the flattest curve is the temperature such that g/kT = 0.1. The other curves (in sequence) represent temperatures where g/kT = 0.8, 1.0, 1.2, 2.0, 4.0 and 10. Note how as the temperature decreases, the sharpness of the population as a function of energy (level) increases. At the low temperature limit, $f(\varepsilon) = 1$ when $\varepsilon < g$, and $f(\varepsilon) = 0$ when $\varepsilon > g$. Another feature to notice is that when $\varepsilon = g$, (i.e., $\gamma = 1$), $f(\varepsilon = g)$ always equals 1/2.

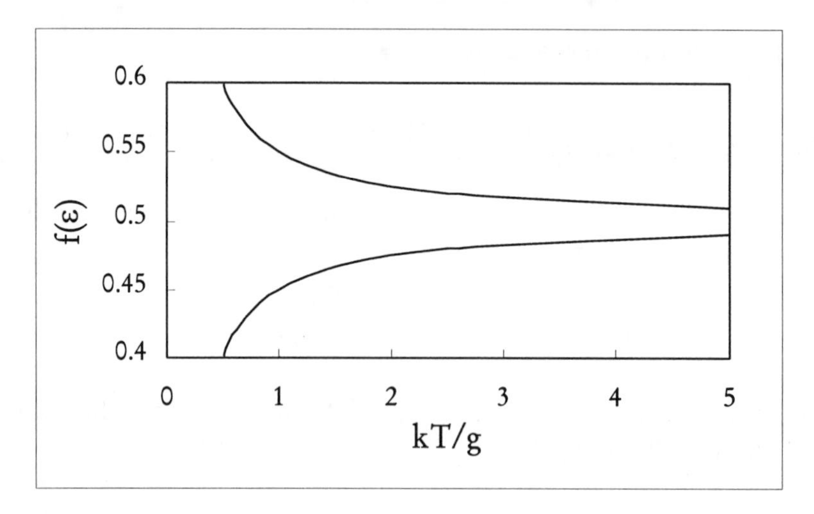

This plot of $f(\varepsilon)$ vs. kT/g for $\gamma = 0.8$ (upper curve) and $\gamma = 1.2$ (lower curve) illustrates that as temperature rises, the population of states with energies above and below the chemical potential (g) all converge to a value of 1/2. As the temperature is lowered, the population of states with energies above the chemical potential decrease and those with energies below the chemical potential increase.

9. Following [I-3] in Appendix I [p. 557 and following], we can find a power series expansion of $(1 - x)^{-1}$ from its derivatives:

$$\frac{d\,(1-x)^{-1}}{dx} = -1(1-x)^{-2}\,\frac{d\,(1-x)}{dx} = -1(1-x)^{-2}(-1) = (1-x)^{-2}.$$

$$\frac{d^2\,(1-x)^{-1}}{dx^2} = \frac{d\,(1-x)^{-2}}{dx} = -2(1-x)^{-3}(-1) = 2(1-x)^{-3}.$$

$$\frac{d^3 (1-x)^{-1}}{dx^3} = \frac{d\, 2(1-x)^{-3}}{dx} = -3(2)(1-x)^{-3}(-1) = 6(1-x)^{-4}.$$

The pattern should be evident by now that the nth derivative of $(1 - x)^{-1}$ is simply $n!(1 - x)^{-(n + 1)}$. Since we're expanding the function about $x = 0$, we need to evaluate the derivatives at $x = 0$:

$$\left.\frac{d^n (1-x)^{-1}}{dx^n}\right|_{x=0} = n!(1-x)^{-(n+1)}\Big|_{x=0} = n!.$$

Now that we know the derivatives at $x = 0$, we can substitute these values into [I-3] to arrive at

$$(1-x)^{-1} = 1 + x \times 1! + \frac{x^2}{2!} \times 2! + \frac{x^3}{3!} \times 3! + \cdots$$

$$= 1 + x + x^2 + x^3 + \cdots.$$

Now, we apply this expansion to the left-hand side of [11-45]:

$$\frac{1}{1 - e^{-\hbar\omega/kT}} = 1 + \left(e^{-\hbar\omega/kT}\right) + \left(- e^{-\hbar\omega/kT}\right)^2 + \left(- e^{-\hbar\omega/kT}\right)^3 + \cdots$$

$$= 1 + e^{-\hbar\omega/kT} + e^{-2\hbar\omega/kT} + e^{-3\hbar\omega/kT} + \cdots = \sum_n e^{-n\hbar\omega/kT}.$$

10. An expression for the average energy due to vibration is derived in Example 11.2.

$$U_{vib} = \langle E_{vib} \rangle = \frac{N\hbar\omega}{2}\left(1 + \frac{2}{1 - e^{-\hbar\omega/kT}}\right)$$

In this problem, $N\hbar\omega = 23$ kJ mol^{-1}, and it is convenient to use the same value of k as in Problem 4, $k = 8.314 \times 10^{-3}$ kJ mol^{-1} K^{-1}.

$$U_{vib}(1\text{ K}) = \frac{23}{2}\left(1 + \frac{2}{1 - e^{-(23\text{ kJ/mol})/(8.314 \times 10^{-3}\text{ kJ/(mol K)} \times 1\text{ K})}}\right) = 34.5\text{ kJ mol}^{-1}$$

$$U_{vib}(10\text{ K}) = \frac{23}{2}\left(1 + \frac{2}{1 - e^{-(23\text{ kJ/mol})/(8.314 \times 10^{-3}\text{ kJ/(mol K)} \times 10\text{ K})}}\right) = 34.5\text{ kJ mol}^{-1}$$

$$U_{vib}(100\text{ K}) = \frac{23}{2}\left(1 + \frac{2}{1 - e^{-(23\text{ kJ/mol})/(8.314 \times 10^{-3}\text{ kJ/(mol K)} \times 100\text{ K})}}\right) = 34.5\text{ kJ mol}^{-1}$$

$$U_{vib}(300\text{ K}) = \frac{23}{2}\left(1 + \frac{2}{1 - e^{-(23\text{ kJ/mol})/(8.314 \times 10^{-3}\text{ kJ/(mol K)} \times 300\text{ K})}}\right) = 34.502\text{ kJ mol}^{-1}$$

$$U_{vib}(1000\text{ K}) = \frac{23}{2}\left(1 + \frac{2}{1 - e^{-(23\text{ kJ/mol})/(8.314 \times 10^{-3}\text{ kJ/(mol K)} \times 1000\text{ K})}}\right) = 36.043\text{ kJ mol}^{-1}.$$

11. To calculate the average dipole moment in this case requires an explicit summation over the vibrational states:

$$\langle\mu\rangle = \sum_i P_i\, \mu_i = \sum_i \frac{e^{-\omega_i/kT}\mu_i}{q}.$$

For these real (not harmonic) oscillators, we must also find the partition function by explicit summation:

$$q = \sum_i e^{-\omega_i/kT}.$$

At T = 10 K, kT = 0.6950 $\text{cm}^{-1}\ \text{K}^{-1} \times 10\ \text{K} = 6.950\ \text{cm}^{-1}$.

$$\begin{aligned} q(10\text{ K}) &= e^{-2000\text{ cm}^{-1}/6.950\text{ cm}^{-1}} + e^{-5960\text{ cm}^{-1}/6.950\text{ cm}^{-1}} \\ &\quad + e^{-9860\text{ cm}^{-1}/6.950\text{ cm}^{-1}} + e^{-13,700\text{ cm}^{-1}/6.950\text{ cm}^{-1}} \\ &= e^{-287.8} + e^{-857.6} + e^{-1418.7} + e^{-1971.2} \\ &= e^{-287.8}(1 + e^{-569.9} + e^{-1130.9} + e^{-1683.4}). \end{aligned}$$

$$\langle \mu \rangle (10\text{ K}) = \frac{(e^{-287.8} \times 2.91) + (e^{-857.6} \times 3.01) + (e^{-1418.7} \times 3.09) + (e^{-1971.2} \times 3.15)}{e^{-287.8}(1 + e^{-569.9} + e^{-1130.9} + e^{-1683.4})}$$

$$= 2.91\text{ D}.$$

At T = 100 K, kT = 0.6950 cm^{-1} K^{-1} × 100 K = 69.50 cm^{-1}.

$$q(100\text{ K}) = e^{-2000\text{ cm}^{-1}/69.50\text{ cm}^{-1}} + e^{-5960\text{ cm}^{-1}/69.50\text{ cm}^{-1}}$$

$$+ e^{-9860\text{ cm}^{-1}/69.50\text{ cm}^{-1}} + e^{-13,700\text{ cm}^{-1}/69.50\text{ cm}^{-1}}$$

$$= e^{-28.78} + e^{-85.76} + e^{-141.87} + e^{-197.12}$$

$$= 3.170 \times 10^{-13} + 5.687 \times 10^{-38} + 2.436 \times 10^{-62} + 2.465 \times 10^{-82}$$

$$= 3.170 \times 10^{-13}.$$

$$\langle \mu \rangle (100\text{ K}) = [(3.170 \times 10^{-13} \times 2.91) + (5.687 \times 10^{-38} \times 3.01) +$$

$$+ (2.436 \times 10^{-62} \times 3.09) + (2.465 \times 10^{-82} \times 3.15)] / 3.170 \times 10^{-13}$$

$$= 2.91\text{ D}.$$

At T = 300 K, kT = 0.6950 cm^{-1} K^{-1} × 300 K = 208.5 cm^{-1}.

$$q(300\text{ K}) = e^{-2000\text{ cm}^{-1}/208.5\text{ cm}^{-1}} + e^{-5960\text{ cm}^{-1}/208.5\text{ cm}^{-1}}$$

$$+ e^{-9860\text{ cm}^{-1}/208.5\text{ cm}^{-1}} + e^{-13,700\text{cm}^{-1}/208.5\text{ cm}^{-1}}$$

$$= e^{-9.592} + e^{-28.59} + e^{-47.29} + e^{-65.71}$$

$$= 6.827 \times 10^{-5} + 3.833 \times 10^{-13} + 2.899 \times 10^{-21} + 2.901 \times 10^{-29}$$

$$= 6.827 \times 10^{-5}.$$

$$\langle \mu \rangle (300\text{ K}) = [(6.827 \times 10^{-5} \times 2.91) + (3.833 \times 10^{-13} \times 3.01) +$$

$$+ (2.899 \times 10^{-21} \times 3.09) + (2.901 \times 10^{-29} \times 3.15)] / 6.827 \times 10^{-5}$$

$$= 2.91\text{ D}.$$

At T = 1000 K, kT = 0.6950 cm^{-1} K^{-1} × 1000 K = 695.0 cm^{-1}.

$$q(1000\text{ K}) = e^{-2000\text{ cm}^{-1}/695.0\text{ cm}^{-1}} + e^{-5960\text{ cm}^{-1}/695.0\text{ cm}^{-1}}$$

$$+ e^{-9860\text{ cm}^{-1}/695.0\text{ cm}^{-1}} + e^{-13{,}700\text{ cm}^{-1}/695.0\text{ cm}^{-1}}$$

$$= e^{-2.878} + e^{-8.576} + e^{-14.19} + e^{-19.71}$$

$$= 0.05625 + 0.00019 + 6.876 \times 10^{-7} + 2.755 \times 10^{-9}$$

$$= 0.05644.$$

$$\langle \mu \rangle (1000\text{ K}) = [(0.05625 \times 2.91) + (0.00019 \times 3.01) +$$

$$+ (6.876 \times 10^{-7} \times 3.09) + (2.755 \times 10^{-9} \times 3.15)] / 0.05644$$

$$= 2.9103\text{ D}.$$

12. For this problem, we first use [11-46] to find $q_{vibrational}$ for the harmonic oscillator. However, to account for that anharmonic term, we will have to calculate $q_{vibrational}$ by explicit summation (as in the previous problem).

$$q_{vibrational} = \sum_i e^{-E_i/kT}$$

where $E_i = 23\text{ kJ mol}^{-1}(i + 1/2) - 0.1\text{ kJ mol}^{-1}(i + 1/2)^2$.

The number of computations required makes this well suited for solving with a spreadsheet program. We enter the energy expression above and calculate enough terms for $q_{vibrational}$ to converge ($i = 6,7$ should be sufficient; we check that the last term is less than 10^{-5} times the sum of all the terms to ensure convergence).

The table compares the harmonic and anharmonic vibrational partition functions. Note that the anharmonicity has little effect over the entire range of temperatures examined.

The plot of these results follow the table. The heavy, broken line is the harmonic partition function, and the light, solid line is the anharmonic partition function.

T (K)	q (harmonic)	q(anharmonic)
10	8.47E-61	1.14E-60
30	9.46E-21	1.05E-20
50	9.67E-13	1.03E-12
70	2.62E-09	2.73E-09
90	2.12E-07	2.19E-07
100	9.84E-07	1.01E-06
150	9.89E-05	0.000101
200	0.000992	0.001007
250	0.003955	0.004003
300	0.009946	0.010046
350	0.019223	0.019389
400	0.031523	0.031763
450	0.046344	0.046661
500	0.063137	0.06353
600	0.100726	0.101274
700	0.141337	0.142045
800	0.18323	0.184109
900	0.225474	0.226541
1000	0.267601	0.268877

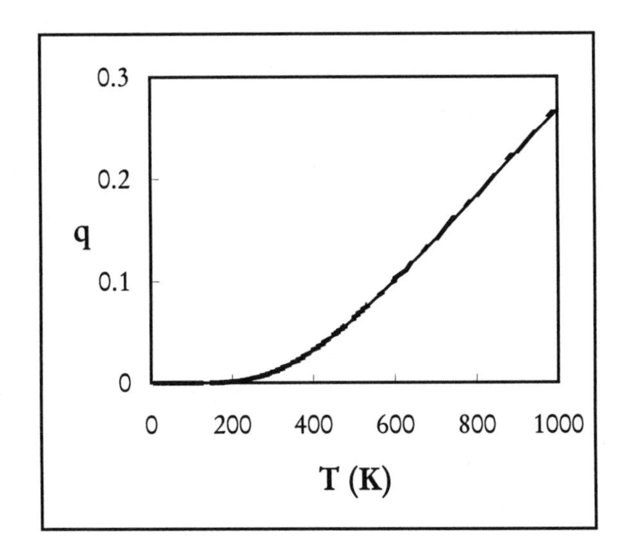

13. Using the high-temperature approximation [11-49], we obtain $q_{rot} = kT/B$. Equation [11-48] is the exact partition function, which must be calculated by explicit summation. Again, this is best solved using a spreadsheet program to evaluate the terms and carry out the summation.

	T = 100 K		T = 300 K	
	[11-48]	[11-49]	[11-48]	[11-49]
$B = 0.2\ cm^{-1}$:	347.8 (76)	347.5	1042.83 (134)	1042.5
$B = 2.0\ cm^{-1}$:	35.09 (24)	34.75	104.58 (42)	104.25
$B = 20\ cm^{-1}$:	3.828 (8)	3.475	10.764 (13)	10.425

The numbers in parentheses represent the number of terms necessary to converge the summation.

14. The exact expression for the average energy of a collection of harmonic oscillators is

$$U_{vib} = \frac{N\hbar\omega}{2}\left(1 + \frac{2}{1 - e^{-\hbar\omega/kT}}\right).$$

The power series expansion for e^x was derived in Problem 18 in Chapter 2. In Example 11.2, the series was truncated after the linear term. Now, we include the second- and third order terms.

$$e^{-\hbar\omega/kT} \approx 1 + \frac{-\hbar\omega}{kT} + \frac{1}{2}\left(\frac{-\hbar\omega}{kT}\right)^2 + \frac{1}{6}\left(\frac{-\hbar\omega}{kT}\right)^3.$$

Substitute this back into the exact energy expression.

$$U_{vib} \approx \frac{N\hbar\omega}{2}\left(1 + \frac{2}{1 - \left[1 - \frac{\hbar\omega}{kT} + \frac{1}{2}\left(\frac{\hbar\omega}{kT}\right)^2 - \frac{1}{6}\left(\frac{\hbar\omega}{kT}\right)^3\right]}\right)$$

$$U_{vib} \approx \frac{N\hbar\omega}{2}\left(1+\frac{2}{\frac{\hbar\omega}{kT}-\frac{1}{2}\left(\frac{\hbar\omega}{kT}\right)^2+\frac{1}{6}\left(\frac{\hbar\omega}{kT}\right)^3}\right)$$

$$U_{vib} \approx \frac{N\hbar\omega}{2}\left(1+\frac{2kT}{\hbar\omega\left[1-\frac{1}{2}\left(\frac{\hbar\omega}{kT}\right)+\frac{1}{6}\left(\frac{\hbar\omega}{kT}\right)^2\right]}\right)$$

$$U_{vib} \approx \frac{N\hbar\omega}{2}+\frac{NkT}{\left[1-\frac{1}{2}\left(\frac{\hbar\omega}{kT}\right)+\frac{1}{6}\left(\frac{\hbar\omega}{kT}\right)^2\right]}.$$

We can see the effect of including the additional two terms in the expansion of the exponential; in the example the approximation to U was simply

$$U_{vib} \approx \frac{N\hbar\omega}{2}+NkT.$$

Let's examine the difference these terms make in a specific example. First, the approximation developed in the example predicts

$$U_{vib} = 23/2 \text{ kJ mol}^{-1} + 8.314 \times 10^{-3} \text{ kJ mol}^{-1} \times 300 \text{ K}$$

$$= 11.5 + 2.494 = 13.994 \text{ kJ mol}^{-1}.$$

Here, $\hbar\omega = 23$ kJ mol^{-1}, and when we set T = 300 K:

$$\frac{\hbar\omega}{kT} = \frac{23 \text{ kJ mol}^{-1}}{8.314 \times 10^{-3} \text{ kJ mol}^{-1} \text{ K}^{-1} \times 300 \text{ K}} = 9.2214.$$

The denominator from our approximate expression above is then

$$1-\frac{1}{2}\left(\frac{\hbar\omega}{kT}\right)+\frac{1}{6}\left(\frac{\hbar\omega}{kT}\right)^2=1-\frac{9.2214}{2}+\frac{9.2214^2}{6}=10.56.$$

This leads to the following value of the vibrational energy:

$$U_{vib} = 11.5 + 2.494/10.56 = 11.736.$$

$$\text{Percent error} = \frac{11.736-13.994}{11.736}\times 100 = -19.2\%.$$

15. Equation [11-29] relates g, the molecular chemical potential, to q, the molecular partition coefficient:

$$e^{-g/kT} = q/N.$$

Taking the natural log of both sides and rearranging leads to

$$g = -kT \ln (q / N).$$

The Gibbs' free energy, G, is simply Ng, so finally,

$$G = -NkT \ln (q / N).$$

16. The heat capacity of an ideal gas of rotating, harmonically vibrating diatomic molecules is given in [11-62]:

$$C_v = \frac{5}{2}Nk + Nk\left(\frac{\hbar\omega}{kT}\right)^2 \frac{e^{-\hbar\omega/kT}}{\left(1-e^{-\hbar\omega/kT}\right)^2}.$$

$Nk = R = 8.314$ J mol^{-1} K^{-1}.

$\hbar\omega/k = 23$ kJ mol^{-1}/8.314 × 10^{-3} kJ mol^{-1} K^{-1} = 2766.4 K^{-1}.

$$C_v = 8.314 \text{ J mol}^{-1}\text{K}^{-1}\left(\frac{5}{2}+\left(\frac{2766.4}{T}\right)^2 \frac{e^{-2766.4/T}}{\left(1-e^{-2766.4/T}\right)^2}\right).$$

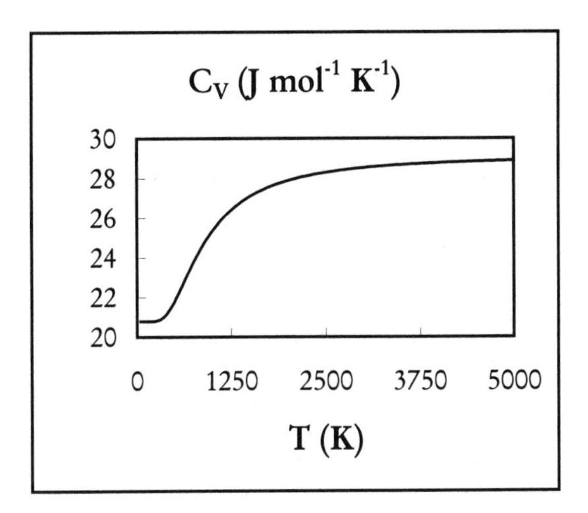

17. The equipartition principle says that each degree of freedom that enters the Hamiltonian quadratically contributes a factor of NkT/2 to the internal energy. Translational and rotational motions, which have no potential energy term, contribute just NkT/2; vibrational motions, which have both kinetic and potential energy terms that are quadratic, contribute NkT.

    Acetylene, C_2H_2, possesses 12 degrees of freedom (4 atoms × 3 coordinates for each atom); since it is linear, it has only two rotations and three translations, leaving seven vibrational motions. U = 3NkT/2 (translation) + NkT (rotation) + 7NkT (vibration) = 19NkT/2.

    Diimide, N_2H_2, possesses 12 degrees of freedom, the same as acetylene, except that it is not linear. Therefore, there are 6 (12 – 6) vibrations, unlike acetylene, which had 7. U = 3NkT/2 + 3NkT/2 + 6NkT = 9NkT.

    Glyoxal, $C_2H_2O_2$, is also nonlinear. The number of vibrations is 12 (18 – 6), so U = 3NkT/2 + 3NkT/2 + 12 NkT = 15NkT.

Chapter 12

Magnetic Resonance Spectroscopy

Exercises

1. Because the nuclear spin states are independent of one another, the total number of states is the product of the spin multiplicities of the individual nuclei.

   For H_2O, $I_H = 1/2$ and $I_O = 0$ [see Appendix V: Table of Atomic Masses and Nuclear Spins], the total number of nuclear spin states, N, is

   $$N = (2(1/2) + 1) \times (2(1/2) + 1) \times (2(0) + 1) = 2 \times 2 \times 1 = 4.$$

   For D_2O, $I_D = 1$, so

   $$N = (2(1) + 1) \times (2(1) + 1) \times (2(0) + 1) = 3 \times 3 \times 1 = 9.$$

   For $D_2{}^{17}O$, $I_{^{17}O} = 5/2$.

   $$N = (2(3/2) + 1) \times (2(3/2) + 1) \times (2(5/2) + 1) = 3 \times 3 \times 6 = 54.$$

2. Each nitrogen can make two transitions, $-1 \Leftrightarrow 0$ and $0 \Leftrightarrow 1$. There will be one of these transitions for every state of the other nuclei, of which there are 12 ($[2(1) + 1] \times [2(1/2) + 1] \times [2(1/2) + 1] = 3 \times 2 \times 2$). For both nitrogens, the total number of transitions is $2 \times 2 \times 12 = 48$.

   If we follow the same analysis for the protons, each proton has one transition, $-1/2 \Leftrightarrow 1/2$. There are 18 states of the other nuclei ($3 \times 3 \times 2$). So the total number of proton transitions is 36 ($2 \times 1 \times 18$).

Additional Exercises

3. The spins of H, D, ^{13}C, and C are 1/2, 1, 1/2, and 0. Since the C nucleus has no spin, it does not contribute to the energy levels. From Appendix V, the g_i values are $g_H = 5.584$, $g_{^{13}C} = 1.404$, and $g_D = 0.857$. From [12-11] we see that the energy separation between spin states is proportional to the g_i value, so $\Delta E_H > \Delta E_{^{13}C} > \Delta E_D$.

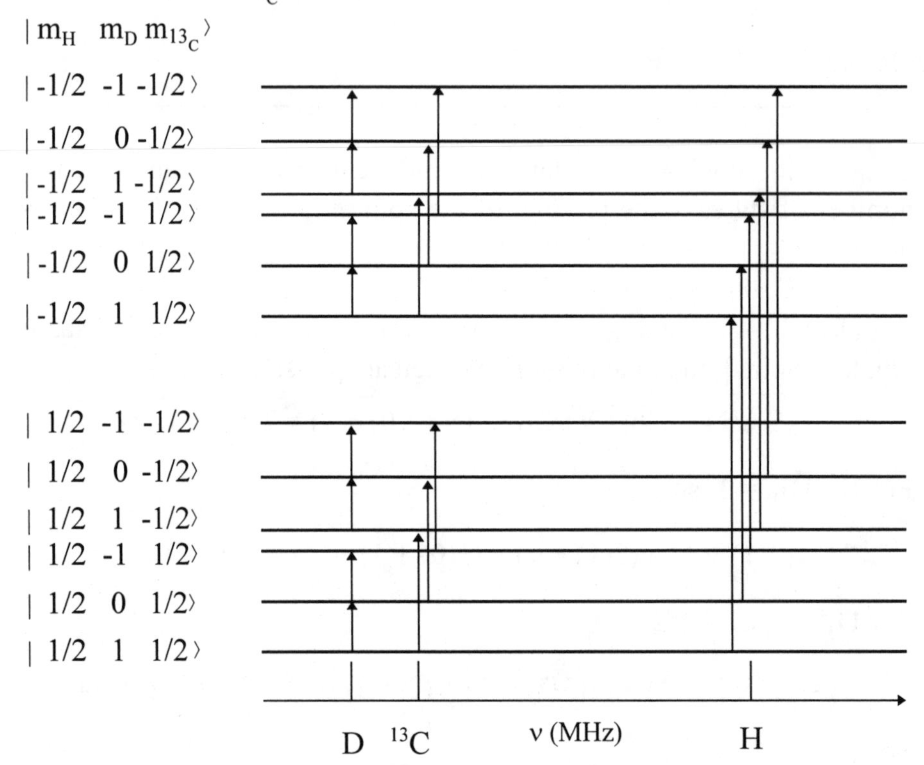

An important feature that this energy diagram shows is the separation of the energy levels in terms of frequency. NMR spectrometers are typically identified by the proton resonance frequency, e.g., a 500 MHz spectrometer. But, for that same spectrometer, the ^{13}C resonance would be about 125 MHz (500 MHz × (1.404/5.584).

4. Appendix III describes two special operators, called the raising and lowering operators, in detail. They are defined as

$$J_+ = J_x + i\,J_y \qquad \text{and} \qquad J_- = J_x - i\,J_y.$$

When they operate on a wavefunction, they raise or lower the z-component of the angular momentum by 1.

$$J_+ \mid J\,M_J \rangle = c \mid J\,M_J + 1 \rangle \text{ and } J_- \mid J\,M_J \rangle = c \mid J\,M_J - 1 \rangle.$$

where c is a constant (see Appendix III for its value).

For this problem, we are to evaluate

$$\left\langle m_{I_1} m_{I_2} | \hat{I}_{x_1}\hat{I}_{x_2} + \hat{I}_{y_1}\hat{I}_{y_2} | m_{I_1} m_{I_2} \right\rangle.$$

[Note: The raising and lowering operators can be formed from any angular momentum operators, not just J. They can be written in terms of the I nuclear angular momentum operators, too.]

$$\hat{I}_{+_1}\hat{I}_{-_2} = \left(\hat{I}_{x_1} + i\hat{I}_{y_1}\right)\left(\hat{I}_{x_2} - i\hat{I}_{y_2}\right) = \hat{I}_{x_1}\hat{I}_{x_2} - i\hat{I}_{x_1}\hat{I}_{y_2} + i\hat{I}_{y_1}\hat{I}_{x_2} + \hat{I}_{y_1}\hat{I}_{y_2}$$

$$\hat{I}_{-_1}\hat{I}_{+_2} = \left(\hat{I}_{x_1} - i\hat{I}_{y_1}\right)\left(\hat{I}_{x_2} + i\hat{I}_{y_2}\right) = \hat{I}_{x_1}\hat{I}_{x_2} + i\hat{I}_{x_1}\hat{I}_{y_2} - i\hat{I}_{y_1}\hat{I}_{x_2} + \hat{I}_{y_1}\hat{I}_{y_2}$$

$$\frac{1}{2}\left(\hat{I}_{+_1}\hat{I}_{-_2} + \hat{I}_{-_1}\hat{I}_{+_2}\right) = \hat{I}_{x_1}\hat{I}_{x_2} + \hat{I}_{y_1}\hat{I}_{y_2}.$$

$$\left\langle m_{I_1} m_{I_2} | \hat{I}_{x_1}\hat{I}_{x_2} + \hat{I}_{y_1}\hat{I}_{y_2} | m_{I_1} m_{I_2} \right\rangle = \left\langle m_{I_1} m_{I_2} | \frac{1}{2}\left(\hat{I}_{+_1}\hat{I}_{-_2} + \hat{I}_{-_1}\hat{I}_{+_2}\right) | m_{I_1} m_{I_2} \right\rangle$$

$$= \frac{1}{2}\left[\left\langle m_{I_1} m_{I_2} | \hat{I}_{+_1}\hat{I}_{-_2} | m_{I_1} m_{I_2} \right\rangle + \left\langle m_{I_1} m_{I_2} | \hat{I}_{-_1}\hat{I}_{+_2} | m_{I_1} m_{I_2} \right\rangle\right]$$

$$= \frac{1}{2}\left[c_1\left\langle m_{I_1} m_{I_2} | m_{I_1} + 1, m_{I_2} - 1\right\rangle + c_2\left\langle m_{I_1} m_{I_2} | m_{I_1} - 1, m_{I_2} + 1\right\rangle\right]$$

= 0, because the functions are orthogonal.

5. a) In formaldehyde, the protons are equivalent, so their spins must first be coupled. $I_{tot} = 1/2 + 1/2, 1/2 - 1/2 = 1,0$. Now, we can make a sketch of the energy levels between the coupled protons and the spin-1/2 ^{17}O.

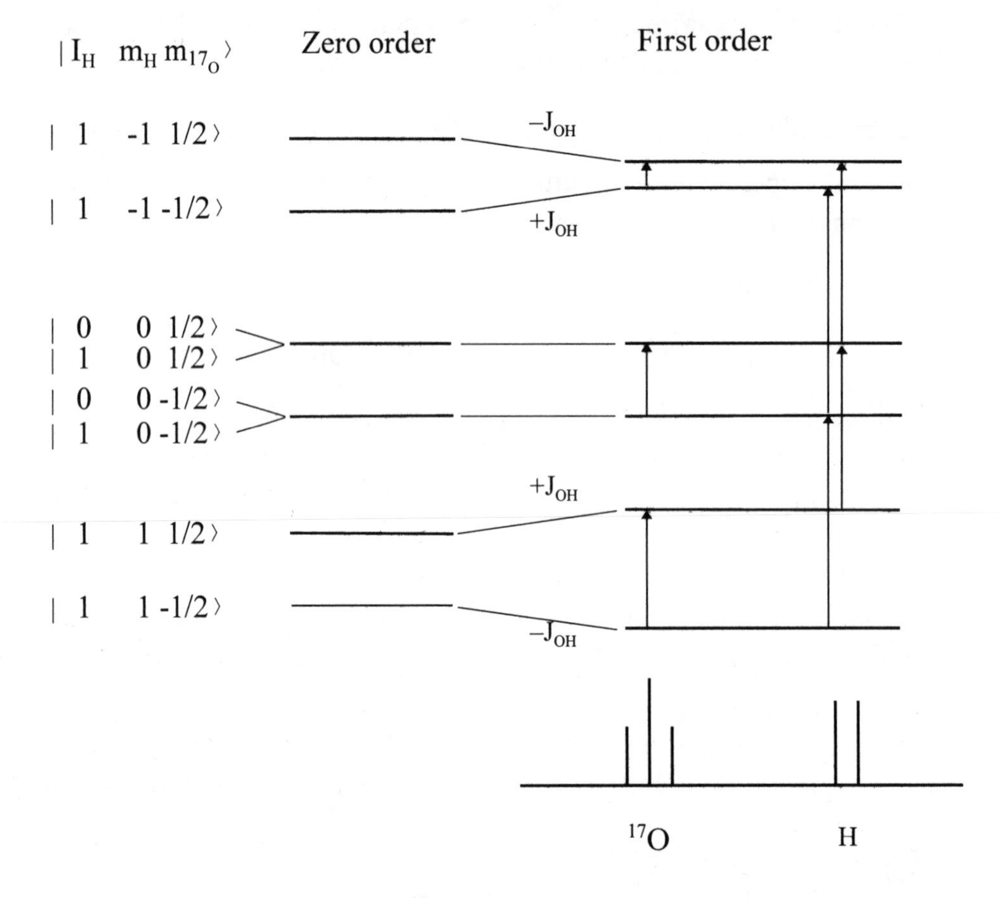

Note that the ordering of the ^{17}O states is reversed with respect to that of H; its g-value is negative. Another key feature is that the first-order splitting is the same for the ^{17}O and H resonances, which can be helpful in identifying particular chemical shifts in a spectrum.

b) For $H_2C^{13}CH_2$ there are two sets of equivalent pairs of hydrogens, the two bonded to the ^{13}C and the other two. These both are coupled as in part a to form $I_H = 1, 0$ states. The ^{13}C is a spin-1/2 particle. The key feature to determining the spectrum of this molecule is the magnitude of the J-couplings. Because of the proximity, the J-coupling between the ^{13}C and the H's bonded to it should be much larger than that of the coupling to the other pair of hydrogens, and the H–H coupling should be smaller yet. Without drawing the complete energy diagram as in part a, we can generate the spectrum by successive splitting.

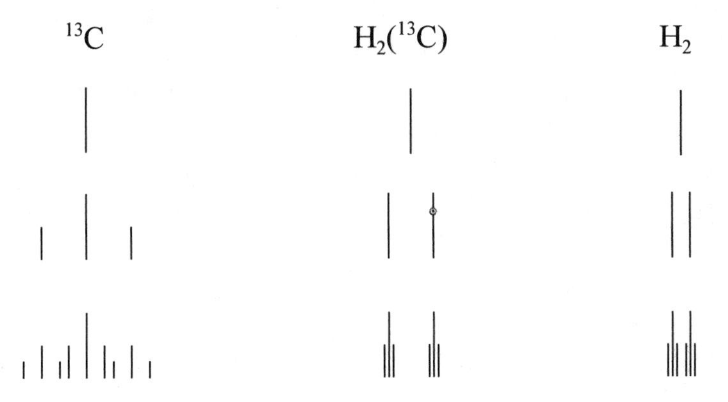

The assumed values for the J-couplings used were $J(^{13}C\text{-}H_2) = 4J(H_2(^{13}C)\text{-}H_2)$ and $J(H_2(^{13}C)\text{-}{}^{13}C) = 10\ J(H_2(^{13}C)\text{-}H_2)$. Note that although the two pairs of hydrogens are qualitatively similar in splitting pattern, the actual values of the splittings distinguish them from one another.

6. The ^{14}N nucleus has a nuclear spin value of 1. The two nitrogens must be coupled (since they are equivalent), so the resulting states will be $I = 2, 1,$ and 0. Consequently, the $I = 2$ state will cause the proton to be split into five equally spaced lines, the $I = 1$ state will cause it to be split into three equally spaced lines, and the $I = 0$ state will not cause splitting. The center lines in each case (due to the $M_I = 0$ state) will overlap, and so the intensity will look like 1:2:3:2:1.

7. The four equivalent protons spins have to be coupled first. The first two protons when coupled will lead to $I = 1, 0$. Coupling to the third proton will result in $I = 3/2, 1/2, 1/2$. Adding the fourth will finally lead to nuclear spin states of $I = 2,1,1,1,0,0$. The $I = 2$ state will cause the inequivalent proton's resonance to be split into five equally spaced lines in the spectrum, the three $I = 1$ states will cause it to be split into three lines, and the two $I = 0$

will not cause any splitting. When these are added together, the final intensity pattern will be 1:4:6:4:1. You may note that the intensity patterns for n-equivalent spin-1/2 particles coupled together are just the binomial coefficients for an nth order polynomial.

8. As for the protons, the deuterons must be successively coupled to determine the splitting patterns. Since I = 1, a single deuteron would lead to a splitting pattern of 1:1:1. Two deuterons coupled together lead to I = 2,1,0. These states lead to splitting into five, three, and one different resonances, respectively, and so when they are combined, the final intensity pattern is 1:2:3:2:1 (as in Problem 6). Coupling to a third deuteron results in I = 3,2,2,1,1,1,0 states. When these contributions are combined, the intensity pattern is 1:3:6:7:6:3:1.

   We can depict this by successively splitting a single resonance by the I = 1 spin of the deuteron.

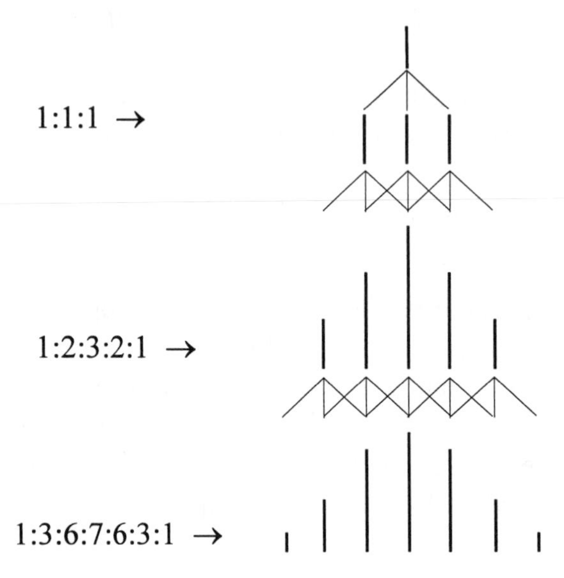

9. For $H_1–{}^{13}C_2≡{}^{13}C_3–{}^{13}C_4≡{}^{13}C_5–H_6$, H_1 and H_6, ${}^{13}C_2$ and ${}^{13}C_5$, and ${}^{13}C_3$ and ${}^{13}C_4$ are each magnetically equivalent. But, in this case, it is not necessary to couple them together to apply first-order perturbation theory, because they are not equivalent when the spin-spin coupling is included. ${}^{13}C_3$ and ${}^{13}C_4$ are degenerate at zero order (they possess the same chemical shifts). However, if the spin-spin coupling between ${}^{13}C_2$ and ${}^{13}C_3$ is included to

first order, the degeneracy is broken (make an energy diagram as in 5a to convince yourself). Then, the spin-spin coupling between $^{13}C_3$ and $^{13}C_4$ can be included, as we've been doing.

The hydrogens and the ^{13}C are all spin-1/2 particles, so the proton will be split into a doublet by the adjacent ^{13}C. $^{13}C_{2(5)}$ will be split into a doublet by the adjacent $^{13}C_{3(4)}$, which will each be further split into another doublet by $H_{1(6)}$. $^{13}C_{3(4)}$ will be split into a doublet by its triply bonded partner, $^{13}C_{2(5)}$, and then each of these will be split into a doublet, as well.

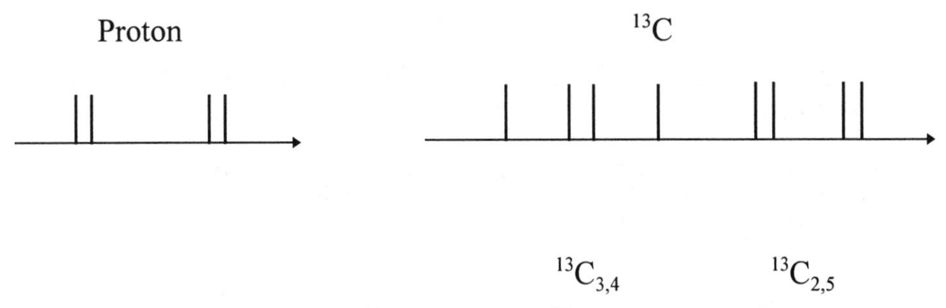

The difference between the $^{13}C_{3,4}$ and the $^{13}C_{2,5}$ results from the assumed smaller spin–spin coupling between the proton and the ^{13}C than between two ^{13}C.

10. For $H^{13}CO\bullet$, the electron spin resonance will be split by the proton and the ^{13}C. For $H^{13}C^{13}C\bullet$, the ESR will be split by the proton and both ^{13}C's.

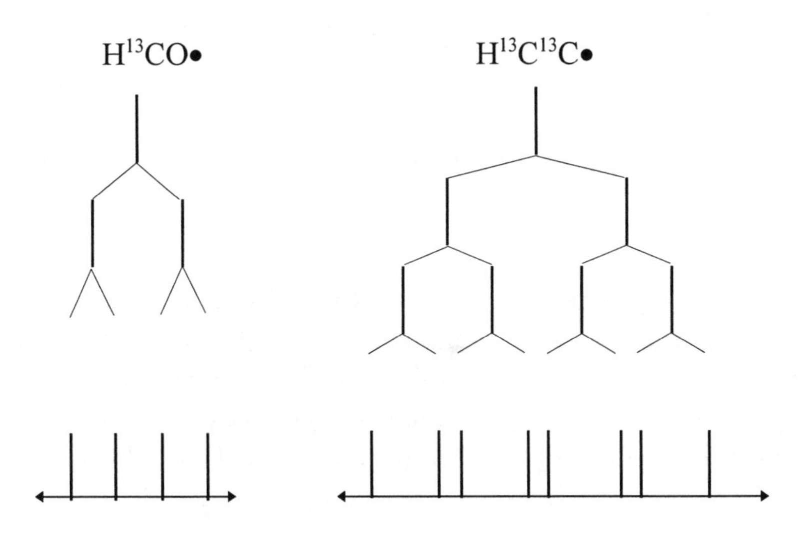

11. $a(CH_2) >> a(CH_3)$ $a(CH_3) >> a(CH_2)$

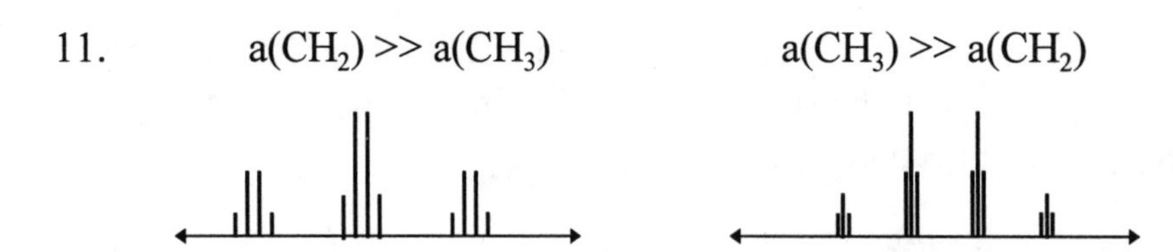

For an intermediate case, let $a(CH_2) = 5$ and $a(CH_3) = 4$

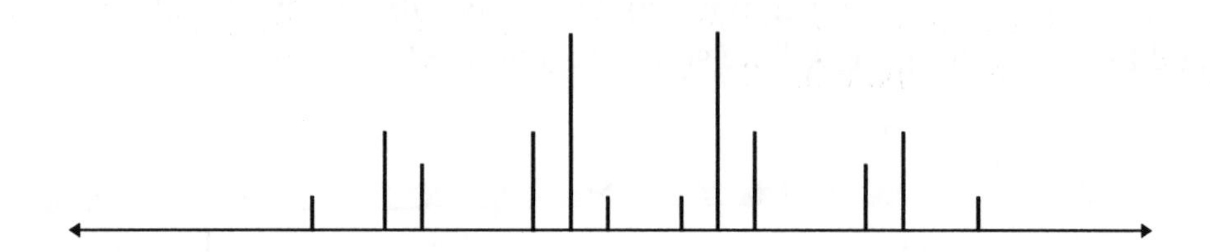

This intermediate case is more likely, where the patterns are not immediately obvious.

12. HOO• $H^{17}OO$• $H^{17}O^{17}O$•

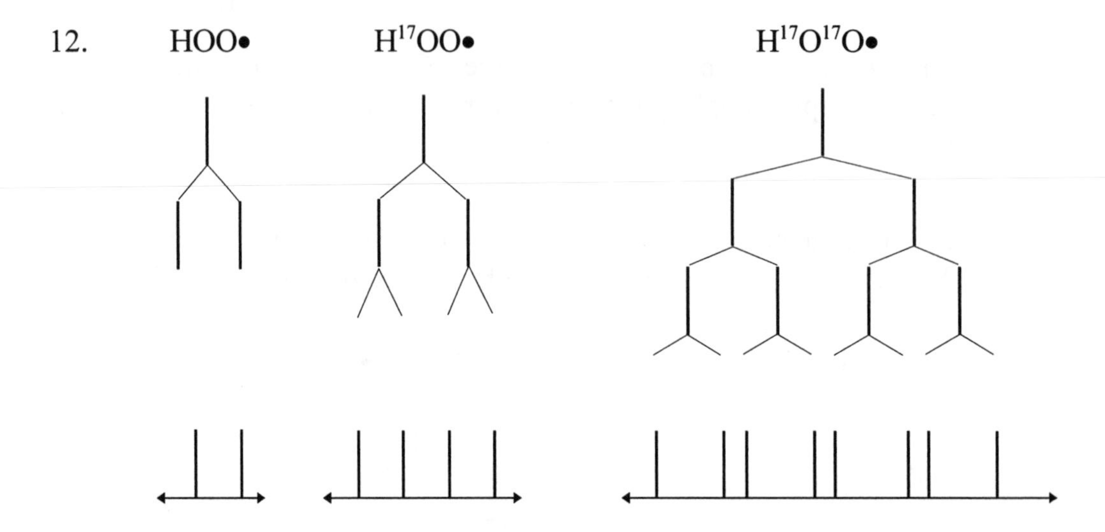

13. This problem concerns two magnetically equivalent nuclei (the acetylenic protons) that experience different magnetic fields. For the magnetic field to have a gradient means that the magnetic field strength varies linearly in a given direction. Let us assign this variation of the magnetic field to lie in the z-direction, and assign a value of H_0 to $z = 0$. Then, $H_z(z) = H_0 + sz$.

Now, let's assign the position of the two protons to be $\pm z_0$. The magnetic field at these positions will then be $H_0 \pm z_0$. If we ignore any spin–spin coupling, the energy difference for the two protons is given by [12-12]

$$\Delta E = \mu_0 H_z g_H (1 - \sigma_H) = \mu_0 [H_0 \pm sz_0] g_H (1 - \sigma_H).$$

Therefore, the two resonances will be split, one to higher frequency and one to lower frequency. The difference in frequency between the two will be proportional to both the size of the field gradient (here designated by s) and the distance separating the two protons ($2z_0$).

Atomic Masses and Percent Natural Abundance of Light Elements

Element	Orbital Occupancy	Isotope Mass (amu)	Natural Abundance	Isotope Mass (amu)	Natural Abundance
1 H	$1s^1$	1.007825	99.985	2.014102	0.015
2 He	$1s^2$	4.002603	100		
3 Li	[He] $2s^1$	6.015121	7.5	7.016003	92.5
4 Be	[He] $2s^2$	9.012182	100		
5 B	[He] $2s^2\ 2p^1$	10.012937	19.9	11.009305	80.1
6 C	[He] $2s^2\ 2p^2$	12.000000	98.90	13.003355	1.10
7 N	[He] $2s^2\ 2p^3$	14.003074	99.63	15.000109	0.37
8 O	[He] $2s^2\ 2p^4$	15.994915	99.762	17.999160	0.200
		16.999131	0.038		
9 F	[He] $2s^2\ 2p^5$	18.998403	100		
10 Ne	[He] $2s^2\ 2p^6$	19.992436	90.48	21.991383	9.25
		20.993843	0.27		
11 Na	[Ne] $3s^1$	22.989768	100		
12 Mg	[Ne] $3s^2$	23.985042	78.99	25.982594	11.01
		24.985837	10.00		
13 Al	[Ne] $3s^2\ 3p^1$	26.981539	100		
14 Si	[Ne] $3s^2\ 3p^2$	27.976927	92.23	29.973771	3.10
		28.976495	4.67		
15 P	[Ne] $3s^2\ 3p^3$	30.973762	100		
16 S	[Ne] $3s^2\ 3p^4$	31.972071	95.02	33.967867	4.21
		32.971459	0.75	35.967081	0.02
17 Cl	[Ne] $3s^2\ 3p^5$	34.968853	75.77	36.965903	24.23
18 Ar	[Ne] $3s^2\ 3p^6$	35.967546	0.337	39.962384	99.600
19 K	[Ar] $4s^1$	38.963707	93.26	40.961835	6.88
		39.963999	0.012		
20 Ca	[Ar] $4s^2$	39.962591	96.94	42.958766	0.14
		41.958618	0.65	43.955481	2.09
31 Ga	[Zn] $4p^1$	68.925580	60.11	70.924701	39.89
32 Ge	[Zn] $4p^2$	69.924250	21.23	73.921177	35.94
		71.922079	27.66	75.921402	7.44
		72.923463	7.73		
33 As	[Zn] $4p^3$	74.921594	100		
34 Se	[Zn] $4p^4$	73.922475	0.89	77.917308	23.78
		75.919212	9.36	79.916520	49.61
		76.919913	7.63	81.916698	8.73
35 Br	[Zn] $4p^5$	78.918336	50.69	80.916289	49.31